Van De Graaff's Photographic Atlas

for the

Zoology Laboratory

SEVENTH EDITION

Byron J. Adams
Brigham Young University

John L. Crawley

925 W. Kenyon Avenue, Unit 12
Englewood, CO 80110

www.morton-pub.com

To the naturalists, environmentalists, and conservation biologists who are dedicated to the preservation of animal species from extinction.

The Galapagos land iguana, *Conolophus subcristatus*, from Isabela Island, is one of two species found in the archipelago. Feral animals have greatly reduced the numbers of land iguanas throughout the Galapagos.

Copyright 1993, 1995, 1998, 2002, 2005, 2009, 2013 by Morton Publishing Company

ISBN13: 978-1-61731-060-7

10 9 8 7 6 5 4 3

All rights reserved. Permission in writing must be obtained from the publisher before any part of this work may be reproduced or transmitted in any form, or by any means, electronic or mechanical, including photocopying and recording or by any information storage or retrieval system.

Printed in the United States of America

Cover Photograph: Galapagos sea lion, *Zalophus wollebaeki*.

Preface

Zoology is an exciting, dynamic, and challenging science. It is the study of organisms within the kingdom Animalia, and it is a fascinating discipline within the broader science of biology. Students are fortunate to be living at a time when insights and discoveries in almost all aspects of zoology are occurring at a very rapid pace. Much of the knowledge learned in a zoology course has application in improving humanity and the quality of life. An understanding of zoology is essential in establishing a secure foundation for more advanced courses in the biological sciences or health sciences.

Zoology is a visually oriented science. *Van De Graaff's Photographic Atlas for the Zoology Laboratory* is intended to provide you with quality photographs of animals, similar to those you may have the opportunity to observe in a zoology laboratory. It is designed to accompany any zoology text or laboratory manual you may be using in the classroom. In certain courses *Van De Graaff's Photographic Atlas for the Zoology Laboratory* could serve as the laboratory manual.

An objective of this atlas is to provide you with a balanced visual representation of the major evolutionary lineages of zoological organisms. Great care has been taken to construct completely labeled, informative figures that are depicted clearly and accurately. Rather than a compilation of striking or rare photos, we have taken great care to ensure that the images in this manual are representative of what you will actually be looking at under the microscope or in the dissection tray in your lab class. The classification schemes, terminology, and phylogenetic trees used in this atlas are as up to date and accurate as possible, and consistent with those of commonly used college zoology texts.

Animals inhabit nearly all aquatic and terrestrial habitats of the biosphere. The greatest number of animals are marine, where the first animals probably evolved. Depending on the classification scheme, animals may be grouped into as many as 40 extant phyla. The most commonly known phylum is Chordata, which includes the subphylum Vertebrata, or the backboned animals. Chordates, however, constitute only about 4.2% of all the described animal species. All other animals are frequently referred to as invertebrates, and account for more than 95% of all animal species.

Several dissections of invertebrate and vertebrate animals were completed and photographed in the preparation of this atlas. An understanding of the structure of an animal is requisite to learning about physiological mechanisms, and even how the animal functions in its environment. The selective pressures that determine evolutionary changes frequently have an influence on anatomical structures. The study of dissected specimens, therefore, provides insight on phylogenetic relationships, or how one group of organisms is related to another.

Some zoology laboratories have the resources to provide students with opportunities for doing selected invertebrate and vertebrate dissections. For these students, the photographs contained in this atlas will be a valuable source for identification of structures on your specimens as they are dissected and studied. If dissection specimens are not available, the excellent photographs of carefully dissected prepared specimens presented throughout this atlas will be an adequate substitute. Care has gone into the preparation of these specimens to depict and identify the principal body structures from representative specimens of each of the major animal phyla. As the anatomy of the various animal specimens is studied in this atlas, note the structural similarities shared from one group to another. These homologous features provide important insights concerning the shared evolutionary history of these animals, and they are particularly interesting as they pertain to our own species.

The information contained in Chapters 1 and 2 is intended to provide you with an orientation to the basic structure of an animal and an understanding of how cells divide. This edition of *A Photographic Atlas for the Zoology Laboratory* contains a discussion of the protists, in Chapter 3. Many zoologists regard protists as sharing common ancestry with multicellular animals. The animal phyla are presented in Chapters 4 through 17. Chapter 18 of this atlas is devoted to the biology of the human animal, which is presented in many zoology textbooks and courses. In that chapter, you are provided with a complete set of photographs for each of the human body systems. Human cadavers have been carefully dissected and photographs taken to clearly depict each of the principal organs from each of the body systems. Selected radiographs (X-rays), CT scans, and MR images depict structures from living persons and thus provide an applied dimension to this portion of the atlas.

Preface to Seventh Edition

The success of the previous editions of *Van De Graaff's Photographic Atlas for the Zoology Laboratory* provided opportunities to make changes to enhance the value of this new edition in aiding students in learning about animals. The revision of this atlas presented in its seventh edition required planning, organization, and significant work. As authors we have the opportunity and obligation to listen to the critiques and suggestions from students and faculty who have used this atlas. This constructive input is appreciated and has resulted in a greatly improved atlas.

One objective in preparing this edition of the atlas was to create an inviting pedagogy. The page layout was improved by careful selection of updated, new, and replacement photographs. All new illustrations were added, including key cladograms making the connections between taxonomy, morphology, and evolutionary history more intuitive. Each image in this atlas was carefully evaluated for its quality, effectiveness, and accuracy. Quality photographs of detailed dissections were updated enhancing the value of this edition. Reformatting of the pedagogy enabled more photographs, photomicrographs, enlarged images in certain chapters, and additional photographs of representative animals. Micrographs were chosen that would closely approximate what students would see in the lab. Perhaps most important to this seventh edition was Dr. Byron Adams, Brigham Young University. Byron has brought important professional input, and rounded out the team.

About the Authors

Byron J. Adams

Byron grew up on a small farm in rural northeastern California, where his parents and schoolteachers nurtured his love of the natural world. He completed his undergraduate degree in Zoology in 1993 from Brigham Young University with an emphasis in marine biology and his Ph.D. in Biological Sciences from the University of Nebraska in 1998. Following a short stint as a postdoctoral fellow at the University of California-Davis, Byron took his first faculty position at the University of Florida prior to returning to Brigham Young University.

Byron's approach to understanding biology involves inferring evolutionary processes from patterns in nature. His research programs in biodiversity, evolution, and ecology have had the continuous support of the National Science Foundation as well as other agencies, including the United States Department of Agriculture and the National Human Genome Research Institute. His most recent projects involve fieldwork in Antarctica, where he and his colleagues are studying the relationship between biodiversity, ecosystem functioning, and climate change. When he's not freezing his butt off in the McMurdo Dry Valleys or southern Transantarctic Mountains, he makes his home in Woodland Hills, Utah.

John L. Crawley

John spent his early years growing up in Southern California, where he took every opportunity to explore nature and the outdoors. He currently resides in Provo, Utah, where he enjoys the proximity to the mountains, desert, and local rivers and lakes.

He received his degree in Zoology from Brigham Young University in 1988. While working as a researcher for the National Forest Service and Utah Division of Wildlife Resources in the early 1990s, John was invited to work on his first project for Morton Publishing, *A Photographic Atlas for the Anatomy and Physiology Laboratory*. After completion of that title John started work on *A Photographic Atlas for the Zoology Laboratory*. To date John has completed five titles with Morton Publishing.

John has spent much of his life observing nature and taking pictures. His photography has provided the opportunity for him to travel widely, allowing him to observe and learn about other cultures and lands. His photos have appeared in national ads, magazines, and numerous publications. He has worked for groups such as Delta Airlines, *National Geographic*, Bureau of Land Management, U.S. Forest Service, and many others. His projects with Morton Publishing have been a great fit for his passion for photography and the biological sciences.

Byron on the plane making his way back from the Transantarctic Mountains heading for McMurdo Station.

John snorkeling with green sea turtles in the Galapagos.

Prelude

Scientists work to determine accuracy in understanding the relationship of organisms even when it requires changing established concepts. DNA sequences, developmental pathways, and morphological structures, along with the fossil record and geological dating, are used to recover the evolutionary history of life (phylogeny) and represent this in a hierarchical classification (taxonomy). New methods for generating and analyzing evolutionary hypotheses continue to improve our understanding of phylogenetic relationships. Because classification schemes that reflect phylogenetic relationships have so much more explanatory power than simple lists of organisms, scientists are constantly updating their classification schemes to reflect these advances in knowledge.

In 1758 Carolus Linnaeus, a Swedish naturalist, assigned all known kinds of organisms into two kingdoms—plants and animals. For over two centuries, this dichotomy of plants and animals served biologists well but has been replaced by the hypothesis of shared common ancestry by three major evolutionary lineages (see exhibit 1). This hypothesis is based primarily on DNA sequence data but corroborates numerous other lines of evidence as well.

Exhibit 1 Domains, Kingdoms, and Representative Examples

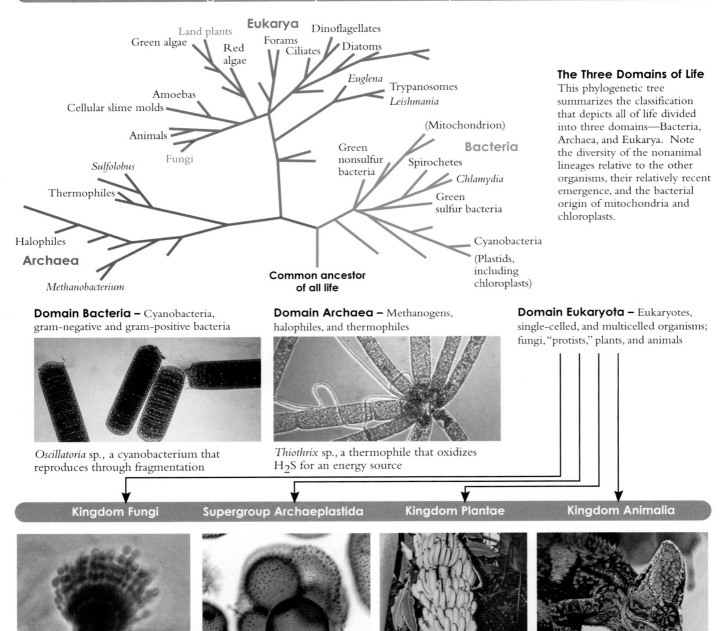

The Three Domains of Life
This phylogenetic tree summarizes the classification that depicts all of life divided into three domains—Bacteria, Archaea, and Eukarya. Note the diversity of the nonanimal lineages relative to the other organisms, their relatively recent emergence, and the bacterial origin of mitochondria and chloroplasts.

Domain Bacteria – Cyanobacteria, gram-negative and gram-positive bacteria

Domain Archaea – Methanogens, halophiles, and thermophiles

Domain Eukaryota – Eukaryotes, single-celled, and multicelled organisms; fungi, "protists," plants, and animals

Oscillatoria sp., a cyanobacterium that reproduces through fragmentation

Thiothrix sp., a thermophile that oxidizes H_2S for an energy source

Aspergillus sp. is a mold that reproduces asexually and sometimes sexually

Volvox sp. is a motile green alga that reproduces asexually or sexually

Musa sp., the banana, is high in nutritional value.

Chamaeleo calyptratus, the veiled chameleon, is known for its ability to change colors according to its mood

Basic Characteristics of Domains

Domain	Characteristics
Domain Bacteria — Bacteria	Prokaryotic cell; single circular chromosome; cell wall containing peptidoglycan; chemosynthetic autotrophs, chlorophyll-based photosynthesis, photosynthetic autotrophs, and heterotrophs; gram-negative and gram-positive forms; lacking nuclear envelope; lacking organelles and cytoskeleton
Domain Archaea — Archaea	Prokaryotic cell; single circular chromosome; cell wall; unique membrane lipids, ribosomes, and RNA sequences; lacking nuclear envelope; some with chlorophyll-based photosynthesis; with organelle, and cytoskeleton
Domain Eukaryota — Eukarya	Single-celled and multicelled organisms; nuclear envelope enclosing more than one linear chromosome; membrane-bound organelles in most; some with chlorophyll-based photosynthesis

Common Classification System of Some Groups of Living Eukaryotes

Eukaryote Supergroups

Excavata — Diplomonads, Parabasalids, and Euglenozoans
Chromalveolata
    Alveolates — Dinoflagellates, Apicomplexans, and Ciliates
    Stramenopiles — Diatoms, Golden algae, Brown algae, and Oomycetes
Rhizaria — Cercozoans, Forams, and Radiolarians
Archaeplastida — Red algae, Green algae, Chlorophytes, Charophytes, and Land plants
Unikonta
    Amoebozoans — Slime molds, Gymnamoebas, and Entamoebas
    Opisthokonts — Nuclearids, Fungi, Choanoflagellates, Animals

* Single-Celled Eukaryote Supergroup Phyla — heterotrophic and phototrophic "protists"
    Phylum Amoebozoa — amoebas and slime molds
    Phylum Heterokontophyta — water molds, diatoms, golden algae
    Phylum Euglenozoa — euglenoids
    Phylum Cryptophyta — cryptomonads
    Phylum Rhodophyta — red algae
    Phylum Dinoflagellata — dinoflagellates
    Phylum Haptophyta — haptophytes

Kingdom Fungi
    Phylum Chytridiomycota — chytrids
    Phylum Zygomycota — zygomycetes
    Phylum Glomeromycota — glomeromycetes
    Phylum Ascomycota — ascomycetes
    Phylum Basidiomycota — basidiomycetes

Kingdom Plantae — bryophytes and vascular plants
    Phylum Hepatophyta — liverworts
    Phylum Anthocerophyta — hornworts
    Phylum Bryophyta — mosses
    Phylum Lycophyta (= Lycopodiophyta) — club moss, ground pines, and spike mosses
    Phylum Pteridophyta — whisk ferns, horsetails, ferns
    Phylum Cycadophyta — cycads
    Phylum Ginkgophyta — Ginkgo
    Phylum Pinophyta (= Coniferophyta) — conifers
    Phylum Gnetophyta — gnetophytes
    Phylum Magnoliophyta (= Anthophyta) — angiosperms (flowering plants)

** Kingdom Animalia — invertebrate and vertebrate animals
    Phylum Ctenophora — comb jellies
    Phylum Porifera — sponges
    Phylum Cnidaria — coral, hydra, and jellyfish
    Phylum Chordata — lancelets, tunicates, and vertebrates
    Phylum Echinodermata — sea stars and sea urchins
    Phylum Hemichordata — acorn worms
    Phylum Nematoda — roundworms
    Phylum Nematomorpha — horsehair worms
    Phylum Tardigrada — water bears
    Phylum Arthropoda — crustaceans, insects, and spiders
    Phylum Kinorhyncha — spiny-crown worms
    Phylum Bryozoa — moss animals
    Phylum Entoprocta — goblet worm
    Phylum Annelida — segmented worms
    Phylum Mollusca — clams, snails, and squids
    Phylum Nemertea — proboscis worms
    Phylum Brachiopoda — lamp shells
    Phylum Phoronida — horseshoe worms
    Phylum Gastrotricha — hairy backs
    Phylum Platyhelminthes — flatworms
    Phylum Rotifera — rotifers

* Historically considered a Kingdom, protists are no longer recognized as such in modern taxonomy. For convenient reference to earlier classification schemes, protist phyla are presented here, but note that each of these is depicted more accurately within the Eukaryote Supergroups.

** Some minor and/or poorly known phyla are not covered in this atlas. Where Phyla are grouped by chapter, they are done so to reflect phylogenetic relationships (with the exception of chapter 3, the unicellular microeukaryotes (protists), and chapter 9, the pseudocoelomates).

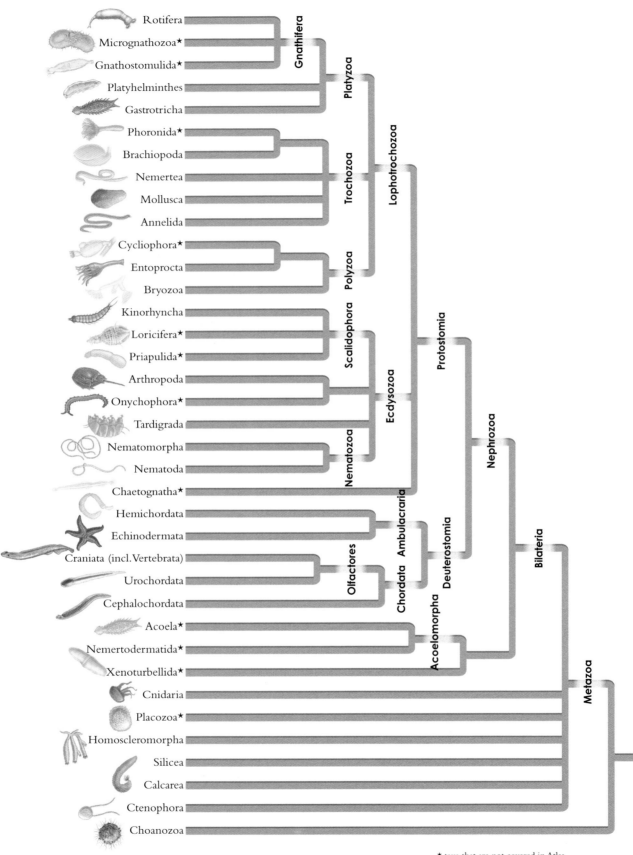

Phylogeny and classification of Metazoa (multicellular animals)

* taxa that are not covered in Atlas

Some Representatives of the Kingdom Animalia

Phylum and Representative Kinds	Characteristics
Porifera: sponges	Multicellular, aquatic animals, with skeletal elements and bodies perforated by pores
Cnidaria: corals, hydra, and jellyfish	Aquatic animals, radially symmetrical, mouth surrounded by tentacles bearing cnidocytes; body composed of epidermis and gastrodermis, separated by mesoglea
Platyhelminthes: flatworms	Elongated, flattened, and bilaterally symmetrical; distinct head containing ganglia; nerve cords; protonephridia or flame cells
Mollusca: clams, snails, and squids	Bilaterally symmetrical with a true coelom, containing a mantle; many have muscular foot and protective shell
Annelida: segmented worms	Body segmented; a series of hearts; hydrostatic skeleton and circular and longitudinal muscles
Nematoda: round worms	Mostly microscopic, unsegmented wormlike; body enclosed in cuticle; whiplike body movement
Arthropoda: crustaceans, insects, and spiders	Body segmented; paired and jointed appendages; chitinous exoskeleton; hemocoel for blood flow
Echinodermata: sea stars and sea urchins	Larvae have bilateral symmetry; adults have pentaradial symmetry; coelom, most contain a complete digestive tract; regeneration of body parts
Chordata: lancelets, tunicates, and vertebrates	Fibrous notochord, pharyngeal gill pouches, dorsal hollow nerve cord, and postanal tail present at some stage in their life cycle

Acknowledgments

Many professionals have assisted in the preparation of *Van De Graaff's Photographic Atlas for the Zoology Laboratory*, seventh edition, and have shared our enthusiasm about its value for students of zoology. The radiographs, CT scans, and MR images have been made possible through the generosity of Gary M. Watts, M.D., and the Department of Radiology at Utah Valley Regional Medical Center. We thank Jake Christiansen, James Barrett, and Austen Slade for their specimen dissections. Others who aided in specimen dissections were Nathan A. Jacobson, D.O., R. Richard Rasmussen, M.D., and Sandra E. Sephton, Ph.D. A special thanks is extended to Kira Wennstrom from Shoreline Community College, Meredith Hamilton from Oklahoma State University, and Bryan Coppedge from Tulsa Community College for their review of the atlas and many helpful suggestions. We are indebted to Douglas Morton and the personnel at Morton Publishing Company for the opportunity, encouragement, and support to prepare this atlas.

Photo Credits

Many of the photographs of living animals were made possible because of the cooperation and generosity of the San Francisco Zoo, San Diego Zoo, San Diego Wild Animal Park, Sea World (San Diego), and Hogle Zoo (Salt Lake City). We are especially appreciative to the professional zoologists at these fine institutions.

Figures 1.15, 3.18, 3.19, 3.20, 3.23, 3.24, and **3.30** from *A Photographic Atlas for the Microbiology Laboratory, 3rd Edition*, by Michael J. Leboffe and Burton E. Pierce. © 2001 Morton Publishing

Figures 5.3, 5.4, 5.5, 5.20, and **8.19** NOAA (National Oceanic and Atmospheric Administration)

Figure 7.36 Ashley Leen, Maine

Figures 7.37, 7.38, and **7.39** from Phillip Colla - Oceanlight.com

Figure 8.20 United States Government - University of California Museum of Paleology

Figure 9.2 (a) NURC/UNCW and NOAA/FGBNMS

Figure 11.19 Ari Pani

Figure 11.21 NOAA Okeanos Explorer Program, INDEX-SATAL 2010

Figures 12.12, 12.15, 12.17, 12.18, 12.19, 13.2, 13.19, and **13.20** from *Comparative Anatomy: Manual of Vertebrate Dissection, 2nd Edition*, by Dale W. Fishbeck and Aurora Sebastiani. © 2008 Morton Publishing

Figure 13.6 Linda Snook, NOAA

Figure 15.17 (a) Louis Porras

Figure 16.3 (k) U.S. Fish and Wildlife Service

Figure 17.3 (c) United States Geological Survey

Figures 17.88, 17.89, 17.90, 17.91, 17.92, 17.93, 17.94, 17.95, 17.96, 17.97, 17.98, 17.99, and **17.100** from *Mammalian Anatomy: The Cat, 2nd Edition*, by Aurora Sebastiani and Dale W. Fishbeck. © 2005 Morton Publishing

We are appreciative of Dr. Wilford M. Hess and Dr. William B. Winborn for their help in obtaining photographs and photomicrographs. The electron micrographs are courtesy of Scott C. Miller. The microscope photo in Figure 1.3 was supplied by Leica Inc.

Book Team

Publisher: Douglas N. Morton
President: David M. Ferguson
Acquisitions Editor: Marta R. Martins
Typography and Text Design: John L. Crawley
Project Manager: Melanie Stafford
Editorial Assistant: Rayna Bailey
Illustrations: Imagineering Media Services, Inc.
Cover Design: Joanne Saliger & Will Kelley

The bobcat, *Felis rufus*, ranges from southern Canada to northern Mexico, including most of the continental United States.

Table of Contents

Chapter 1 - Cells and Tissues	1
Chapter 2 - Perpetuation of Life	13
Chapter 3 - Select Single-Celled Eukaryote Supergroup Phyla ("Protists")	21
Chapter 4 - Porifera	29
Early Metazoa (sponges)	
Chapter 5 - Ctenophora and Cnidaria	33
Radiate animals, including comb jellies, hydra, jellyfish, and corals	
Chapter 6 - Platyhelminthes	43
Acoelomate animals, including flatworms	
Chapter 7 - Mollusca, Brachiopoda, and Bryozoa	51
Mollusks, Brachiopods, and Bryozoa, including clams, snails, squids, lamp shells, and moss animals	
Chapter 8 - Annelida and Nemertea	61
Trochozoan worms, including earthworms, polychaetes, leeches, and proboscis worms	
Chapter 9 - Nematoda and Other Pseudocoelomates	69
Pseudocoelomates, including roundworms	
Chapter 10 - Arthropoda and Tardigrada	75
Panarthropods, including crustaceans, arachnids, insects, and water bears	
Chapter 11 - Echinodermata and Hemichordata	91
Ambulacraria, including sea stars, sea cucumbers, sea urchins, and acorn worms	
Chapter 12 - Chordata	99
Chordates, including amphioxus, fishes, amphibians, reptiles*, birds, and mammals	
Chapter 13 - Chondrichthyes and Osteichthyes: Actinopterygii and Sarcopterygii	107
Fishes, including skates, rays, sharks, and bony fishes	
Chapter 14 - Amphibia	119
Amphibians, including caecilians, salamanders, frogs, and toads	
Chapter 15 - Reptilia* (* = Sauropsida)	131
Reptiles*, including turtles, crocodilians, lizards, and snakes	
Chapter 16 - Aves	145
Birds, including ratite birds and modern birds	
Chapter 17 - Mammalia	155
Mammals, including monotremes, marsupials, and placental mammals	
Chapter 18 - Human Biology	197
Glossary of Terms	225
Index	237

Cells and Tissues

Chapter 1

Animals are heterotrophic organisms that ingest food materials and store carbohydrate reserves as glycogen or fat. The cells of animals lack cell walls, but do contain intercellular connections including desmosomes, gap junctions, and tight junctions. Animal cells are also highly specialized into the specific kinds of tissues depicted in this chapter. Most animals are motile through the contraction of muscle fibers containing actin and myosin proteins. The complex body systems of animals include elaborate sensory and neuromotor specializations that accommodate dynamic behavioral mechanisms.

Cells are the basic structural and functional units of organization within living organisms. A cell is a minute, membrane–enclosed, protoplasmic mass consisting of chromosomes in a nucleus surrounded by cytoplasm containing the specific organelles that function independently but in coordination one with another. Based on structure, there are prokaryotic cells and eukaryotic cells.

Prokaryotic cells lack a membrane-bound nucleus, contain a single bacterial *chromosome* composed of a single strand of *nucleic acid*, contain few organelles, and have a rigid or semirigid *cell wall* outside the *cell (plasma) membrane* that provides shape to the cell. Bacteria are examples of prokaryotic, single-celled, organisms.

Eukaryotic cells contain a nucleus with multiple chromosomes, have numerous specialized *organelles*, and have a differentially permeable *cell (plasma) membrane*. Examples of eukaryotic organisms include protozoa, fungi, algae, plants, and invertebrate and vertebrate animals.

The *nucleus* is the large spheroid body within the eukaryotic cell that contains the *nucleolus, nucleoplasm,* and *chromatin*—the genetic material of the cell. The nucleus is enclosed by a double membrane called the *nuclear membrane,* or *nuclear envelope*. The nucleolus is a dense, nonmembranous body composed of protein and RNA molecules. The chromatin consists of fibers of protein and DNA molecules. Prior to cellular division, the chromatin shortens and coils into rod-shaped *chromosomes*. Chromosomes consist of DNA and proteins called *histones.*

The *cytoplasm* of the eukaryotic cell is the medium of support between the nuclear membrane and the cell membrane. *Organelles* are minute membrane-bound structures within the cytoplasm of a cell that are concerned with specific functions. The cellular functions carried out by the organelles are referred to as *metabolism*. The functions of the principal organelles are listed in Table 1.1. In order for cells to remain alive, metabolize, and maintain *homeostasis,* certain requirements must be met, which include having access to nutrients and oxygen, being able to eliminate wastes, and being maintained in a constant, protective environment.

The *cell membrane* is composed of phospholipid and protein molecules, which gives form to a cell, provides a barrier function, and controls the passage of material into and out of a cell. More specifically, the proteins in the cell membrane provide:

1. structural support;
2. a mechanism of molecule transport across the membrane;
3. enzymatic control of chemical reactions;
4. receptors for water–soluble hormones and other regulatory molecules;
5. cellular markers (antigens), which identify the blood and tissue type.

The phospholipids:

1. repel negative objects due to their negative charge;
2. act as receptors for fat–soluble hormones and other regulatory molecules;
3. form specific cell markers that enable like cells to attach and aggregate into tissues;
4. enter into immune reactions.

Tissues are aggregations of similar cells that perform specific functions. The tissues of the body of a multicellular animal are classified into four principal types, determined by structure and function:

1. *epithelial tissue* covers body and organ surfaces, lines body cavities and lumina (hollow portions of body tubes), and forms various glands;
2. *connective tissue* binds, supports, and protects body parts;
3. *muscle tissue* contracts to produce movements;
4. *nervous tissue* initiates and transmits nerve impulses from one body part to another.

Table 1.1 Structure and Function of Cellular Components

Component	Structure	Function
Cell (plasma) membrane	Composed of protein and phospholipid molecules	Provides form to cell; controls passage of materials into and out of cell
Cytoplasm	Fluid to jelly-like substance	Serves as suspending medium for organelles and dissolved molecules
Endoplasmic reticulum	Interconnecting membrane-lined channels	Enables cell transport and processing of metabolic chemicals
Ribosome	Granules of nucleic acid (RNA) and protein	Synthesizes protein
Mitochondrion	Double-membraned sac with cristae (chambers)	Assembles ATP (cellular respiration)
Golgi complex	Flattened membrane-lined chambers	Synthesize carbohydrates and packages molecules for secretion
Lysosome	Membrane-surrounded sac of enzymes	Digests foreign molecules and worn cells
Centrosome	Mass of protein that may contain rod-like centrioles	Organizes spindle fibers and assists mitosis and meiosis
Vacuole	Membranous sac	Stores and excretes substances within the cytoplasm, regulates cellular turgor pressure
Microfibril and microtubule	Protein strands and tubes	Forms cytoskeleton, supports cytoplasm, and transports materials
Cilium and flagellum	Cytoplasmic extensions from cell; containing microtubules	Movements of particles along cell surface or cell movement
Nucleus	Nuclear envelope (membrane), nucleolus, and chromatin (DNA)	Contains genetic code that directs cell activity; forms ribosomes

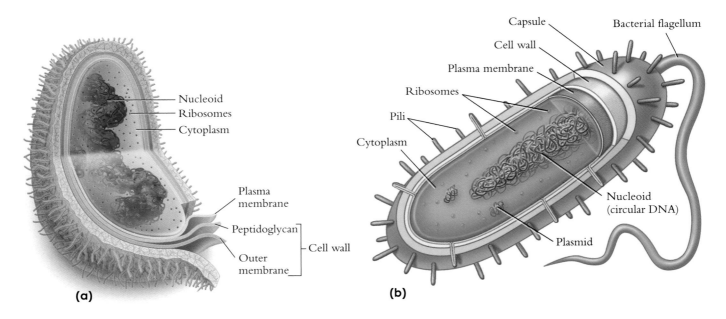

Figure 1.1 Prokaryotic cells, (a) a generalized cell, and (b) a bacterial cell.

Cells and Tissues

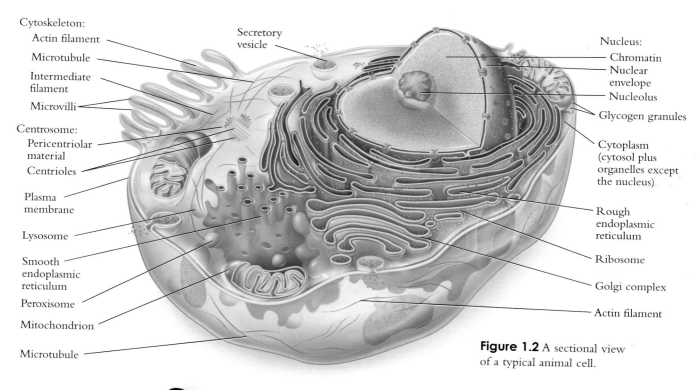

Figure 1.2 A sectional view of a typical animal cell.

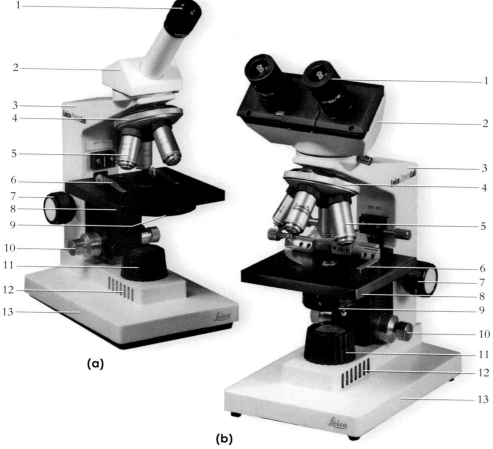

Figure 1.3 (a) A compound monocular microscope, and (b) a compound binocular microscope.
1. Eyepiece (ocular)
2. Head
3. Arm
4. Nosepiece
5. Objective
6. Stage clip
7. Coarse focus adjustment knob
8. Stage
9. Condenser
10. Fine focus adjustment knob
11. Collector lens with iris
12. Illuminator (inside)
13. Base

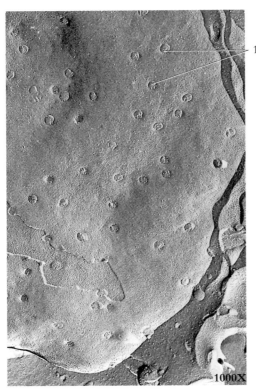

Figure 1.4 An electron micrograph of a freeze-fractured nuclear envelope showing the nuclear pores.
1. Nuclear pores (protein complexes that allow transport of water soluble molecules into the nucleus)

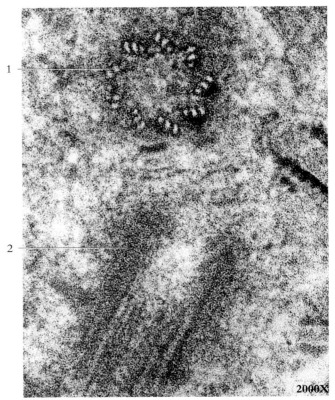

Figure 1.5 An electron micrograph of centrioles. The centrioles are positioned at right angles to one another.
1. Centriole (shown in cross-section)
2. Centriole (shown in longitudinal section)

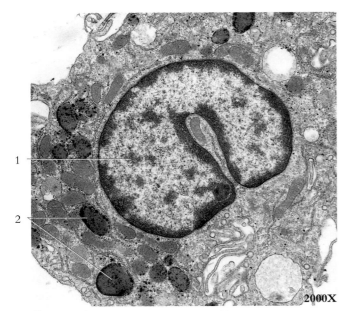

Figure 1.6 An electron micrograph of lysosomes.
1. Nucleus   2. Lysosomes

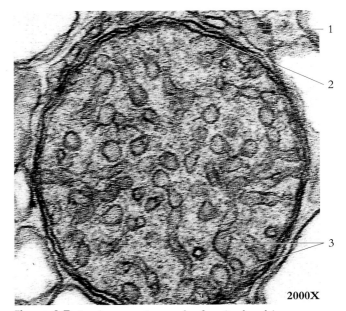

Figure 1.7 An electron micrograph of a mitochondrion (cross-section).
1. Outer membrane   3. Crista
2. Inner membranes

Cells and Tissues

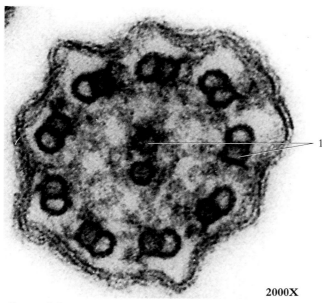

Figure 1.8 An electron micrograph of a cilium (transverse section) showing the characteristic "9 + 2" arrangement of microtubules in the transverse sections.
 1. Microtubules

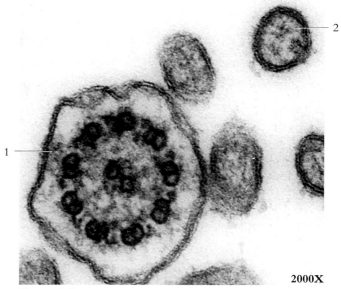

Figure 1.9 An electron micrograph showing the difference between a microvillus and a cilium.
 1. Cilium
 2. Microvillus

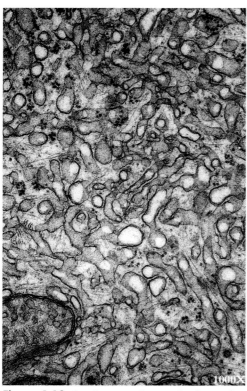

Figure 1.10 An electron micrograph of smooth endoplasmic reticulum from an interstitial cell in the testis.

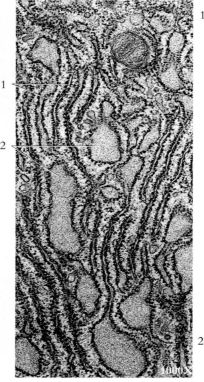

Figure 1.11 An electron micrograph of rough endoplasmic reticulum.
 1. Ribosomes
 2. Cisternae

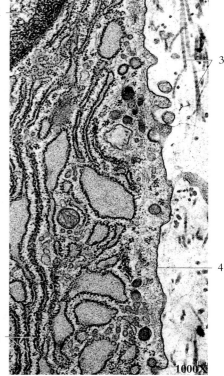

Figure 1.12 Rough endoplasmic reticulum secreting collagenous filaments to the outside of the cell.
 1. Nucleus
 2. Rough endoplasmic reticulum
 3. Collagenous filaments
 4. Cell membrane

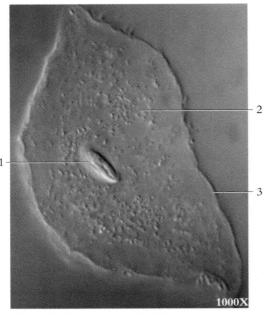

Figure 1.13 An epithelial cell from a cheek scraping.
1. Nucleus
2. Cytoplasm
3. Cell membrane

Figure 1.14 An electron micrograph of an erythrocyte (red blood cell).

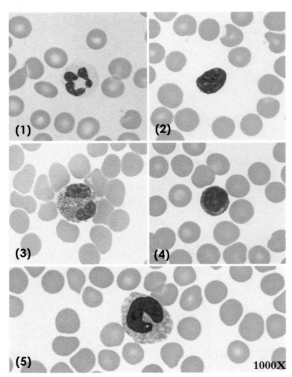

Figure 1.15 Types of leukocytes. Note that each photo contains several erythrocytes; these cells lack nuclei.
1. Neutrophil
2. Basophil
3. Eosinophil
4. Lymphocyte
5. Monocyte

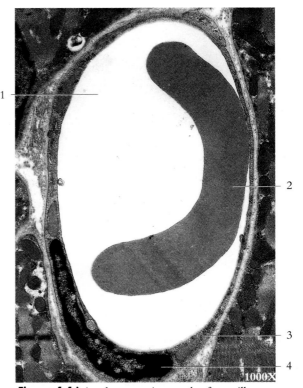

Figure 1.16 An electron micrograph of a capillary containing an erythrocyte.
1. Lumen of capillary
2. Erythrocyte
3. Endothelial cell
4. Nucleus of endothelial cell

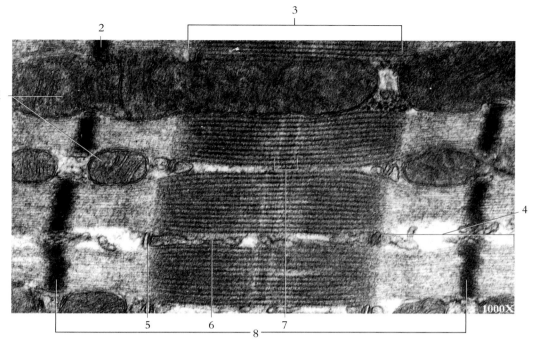

Figure 1.17 An electron micrograph of a skeletal muscle myofibril, showing the striations.

1. Mitochondria
2. Z line
3. A band
4. I band
5. T-tubule
6. Sarcoplasmic reticulum
7. M line
8. Sarcomere

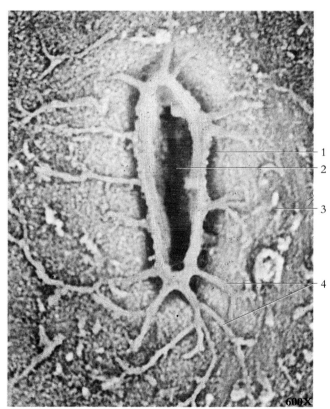

Figure 1.18 An electron micrograph of an osteocyte (bone cell) in cortical bone matrix.

1. Lacuna
2. Osteocyte nucleus
3. Bone matrix
4. Canaliculi

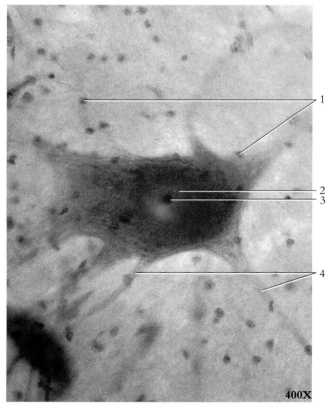

Figure 1.19 An electron micrograph of a neuron smear.

1. Nuclei of surrounding neuroglial cells
2. Nucleus of neuron
3. Nucleolus of neuron
4. Dendrites of neuron

Epithelial Tissue

Epithelial tissue covers the outside of the body and lines all organs. Its primary function is to provide protection.

Simple squamous epithelium

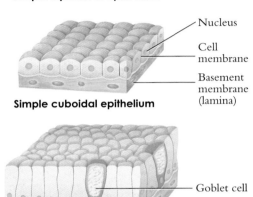
Simple cuboidal epithelium

Simple columnar epithelium

Connective Tissue

Connective tissue functions as a binding and supportive tissue for all other tissues in the organism.

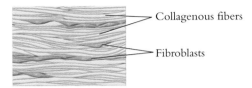
Dense regular connective tissue

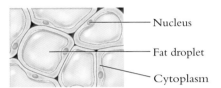
Adipose tissue

Bone tissue

Muscle Tissue

Muscle tissue is a tissue adapted to contract. Muscles provide movement and functionality to the organism.

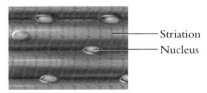
Skeletal muscle

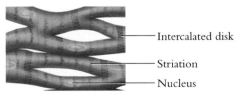
Cardiac muscle

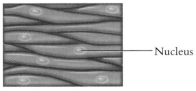
Smooth muscle

Nervous Tissue

Nervous tissue functions to receive stimuli and transmits signals from one part of the organism to another.

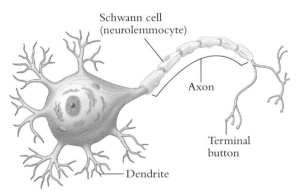
Neuron

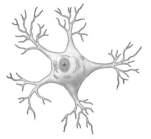
Neurological cell

Figure 1.20 Some examples of animal tissues.

Cells and Tissues

Figure 1.21 Simple squamous epithelium.
1. Single layer of flattened cells

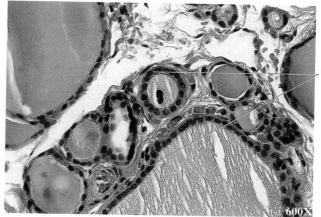

Figure 1.22 Simple cuboidal epithelium.
1. Single layer of cells with round nuclei

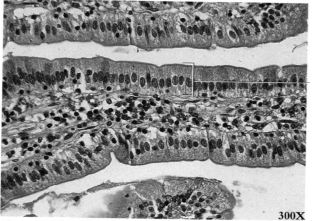

Figure 1.23 Simple columnar epithelium.
1. Single layer of cells with oval nuclei

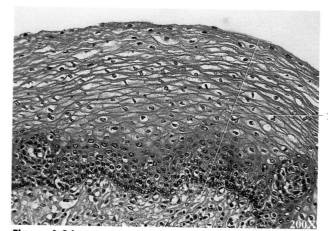

Figure 1.24 Stratified squamous epithelium.
1. Multiple layers of cells, which are flattened at the upper layer

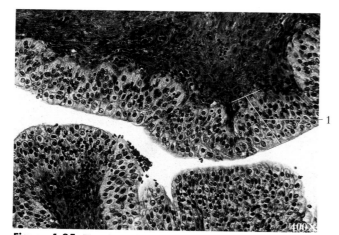

Figure 1.25 Transitional epithelium.
1. Cells are balloon-like at surface

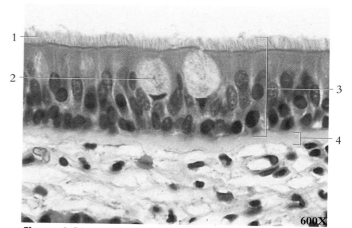

Figure 1.26 Pseudostratified columnar epithelium.
1. Cilia
2. Goblet cell
3. Pseudostratified columnar epithelium
4. Basement membrane

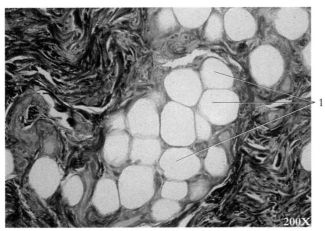

Figure 1.27 Adipose connective tissue.
1. Adipocytes (adipose cells)

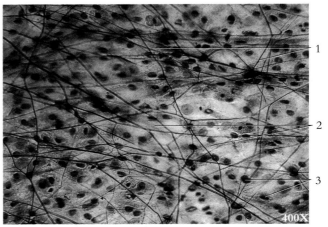

Figure 1.28 Loose connective tissue stained for fibers.
1. Elastic fibers (black)   3. Fibroblasts
2. Collagen fibers (pink)

Figure 1.29 Dense regular connective tissue.
1. Nuclei of fibroblasts arranged in parallel rows between pink-stained collagen fibers

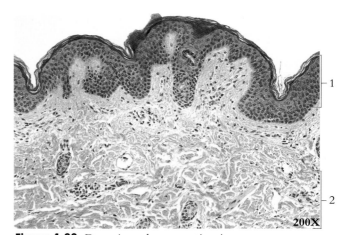

Figure 1.30 Dense irregular connective tissue.
1. Epidermis composed of stratified squamous epithelium
2. Dense irregular connective tissue (reticular layer of dermis)

Figure 1.31 An electron micrograph of dense irregular connective tissue.
1. Collagenous fibers   2. Fibroblasts

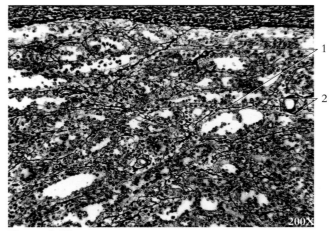

Figure 1.32 Reticular connective tissue.
1. Reticular fibers   2. Lymphocytes

Cells and Tissues

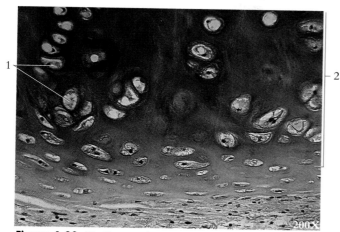

Figure 1.33 Hyaline cartilage.
1. Chondrocytes in lacunae
2. Hyaline cartilage

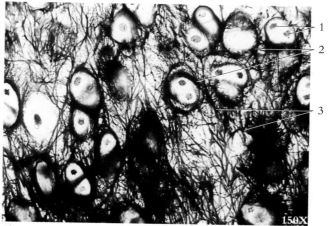

Figure 1.34 Elastic cartilage.
1. Chondrocytes
2. Lacunae
3. Elastic fibers

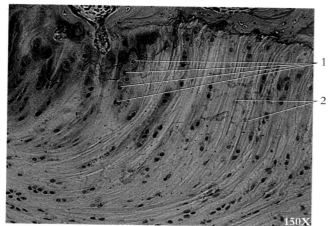

Figure 1.35 Fibrocartilage.
1. Chondrocytes arranged in a row
2. Collagen fibers

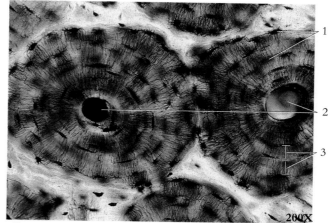

Figure 1.36 A transverse section of two osteons in compact bone tissue.
1. Lacunae containing osteocytes
2. Central (Haversian) canals
3. Lamellae

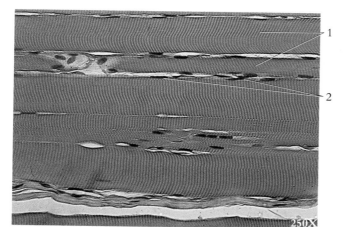

Figure 1.37 A longitudinal section of skeletal muscle tissue.
1. Skeletal muscle cells (note striations)
2. Multiple nuclei in periphery of cell

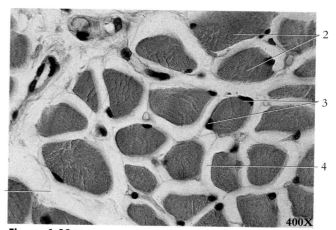

Figure 1.38 A transverse section of skeletal muscle tissue.
1. Perimysium (surrounds bundles of cells)
2. Skeletal muscle cells
3. Multiple nuclei in periphery of cell
4. Endomysium (surrounds individual cells)

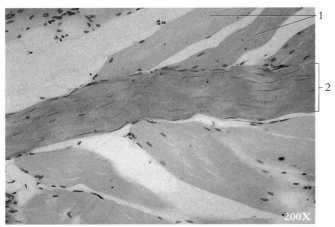

Figure 1.39 An attachment of skeletal muscle to tendon.
1. Skeletal muscle
2. Dense regular connective tissue (tendon)

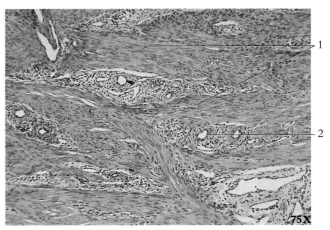

Figure 1.40 Smooth muscle tissue.
1. Smooth muscle
2. Blood vessel

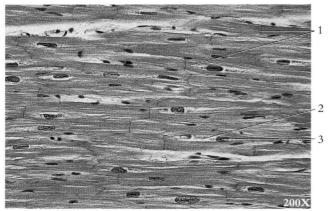

Figure 1.41 Cardiac muscle tissue.
1. Intercalated discs
2. Light-staining perinuclear sarcoplasm
3. Nucleus in center of cell

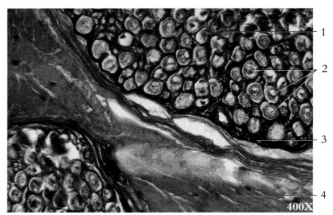

Figure 1.42 A transverse section of a nerve.
1. Endoneurium
2. Axons surrounded by myelin sheath
3. Perineurium
4. Epineurium

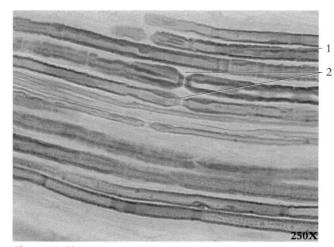

Figure 1.43 A longitudinal preparation of axons.
1. Myelin sheath
2. Neurofibril nodes (nodes of Ranvier)

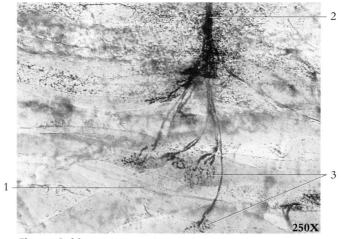

Figure 1.44 A neuromuscular junction.
1. Skeletal muscle fiber
2. Motor nerve
3. Motor end plates

Perpetuation of Life

Chapter 2

The term *cell cycle* refers to how a multicellular organism develops, grows, and maintains and repairs body tissues. In the cell cycle, each new cell receives a complete copy of all genetic information in the parent cell and the cytoplasmic substances and organelles to carry out hereditary instructions.

The animal cell cycle (see fig. 2.3) is divided into: 1) interphase, which includes Gap 1 (G1), Synthesis (S), and Gap 2 (G2) phases; and 2) mitosis, which includes prophase, metaphase, anaphase, and telophase. *Interphase* is the interval between successive cell divisions during which the cell is metabolizing and the chromosomes are directing RNA synthesis. The *G1 phase* is the first growth phase, the *S phase* is when DNA is replicated, and the *G2 phase* is the second growth phase. *Mitosis* (also known as karyokinesis) is the division of the nuclear parts of a cell to form two daughter nuclei with the same number of chromosomes as the original nucleus.

Like the animal cell cycle, the plant cell cycle consists of growth, synthesis, mitosis, and cytokinesis. *Growth* is the increase in cellular mass as the result of metabolism; *synthesis* is the production of DNA and RNA to regulate cellular activity; mitosis is the splitting of the nucleus and the equal separation of the chromatids; and cytokinesis is the division of the cytoplasm that accompanies mitosis.

Unlike animal cells, plant cells have a rigid cell wall that does not cleave during cytokinesis. Instead, a new cell wall is constructed between the daughter cells. Furthermore, many land plants do not have centrioles for the attachment of spindles. The microtubules in these plants form a barrel-shaped anastral spindle at each pole. Mitosis and cytokinesis in plants occur in basically the same sequence as these processes in animal cells.

Asexual reproduction is propagation without sex; that is, the production of new individuals by processes that do not involve *gametes* (sex cells). Asexual reproduction occurs in a variety of microorganisms, fungi, plants, and animals, wherein a single parent produces offspring with characteristics identical to itself. Asexual reproduction is not dependent on the presence of other individuals. No egg or sperm is required. In asexual reproduction, all the offspring are genetically identical (except for mutants). Types of asexual reproduction include:

1. *fission*—subdivision of a cell (or organism, population, species, etc.) into to separate parts. Binary fission produces two separate parts; multiple fission produces more than two separate parts (cells, populations, species, etc.);
2. *sporulation*—multiple fission: many cells are formed and join together in a cystlike structure (protozoans and fungi);
3. *budding*—buds develop organisms like the parent and then detach themselves (hydras, yeast, certain plants); and
4. *fragmentation*—organisms break into two or more parts, and each part is capable of becoming a complete organism (algae, flatworms, echinoderms).

Sexual reproduction is propagation of new organisms through the union of genetic material from two parents. Sexual reproduction usually involves the fusion of haploid gametes (such as sperm and egg cells) during fertilization to form a zygote.

The major biological difference between sexual and asexual reproduction is that sexual reproduction produces genetic variation in the offspring. The combining of genetic material from the gametes produces offspring that are different from either parent and contain new combinations of characteristics. This may increase the ability of the species to survive environmental changes or to reproduce in new habitats. The only genetic variation that can arise in asexual reproduction comes from mutations.

Figure 2.1 Sexual reproduction. A pair of cinnamon teal, *Anas cyanoptera*, in early spring.

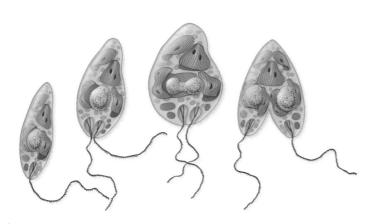

Binary fission

A single cell divides, forming two separate cells. Fission occurs in bacteria, protozoans, and other single-celled organisms.

Figure 2.2 Types of asexual reproduction.

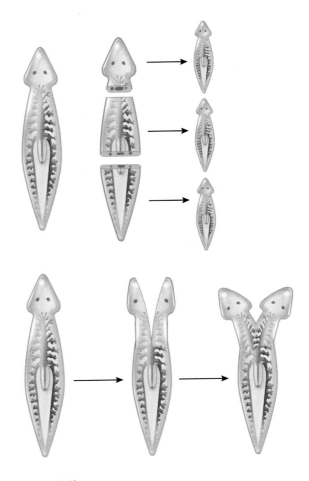

Fragmentation

An organism breaks into two or more parts, each capable of becoming a complete organism. Fragmentation occurs in flatworms and echinoderms.

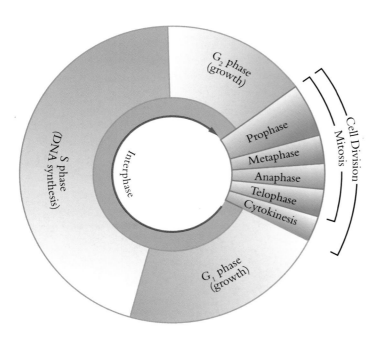

Figure 2.3 An animal cell cycle.

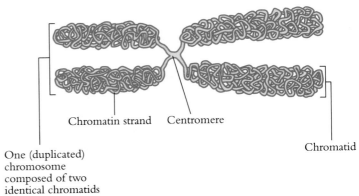

Figure 2.4 Each duplicated chromosome consists of two identical chromatids attached at the centrally located and constricted centromere.

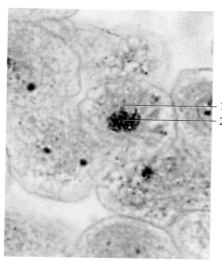

Prophase
Each duplicated chromosome consists of two chromatids joined by a centromere. Spindle fibers extend from each centriole.
1. Aster around centriole
2. Chromosomes

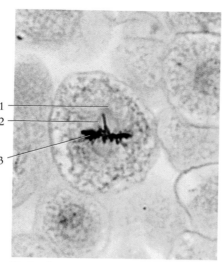

Metaphase
The chromosomes are positioned at the equator. The spindle fibers from each centriole attach to the centromeres.
1. Aster around centriole
2. Spindle fibers
3. Chromosomes at equator

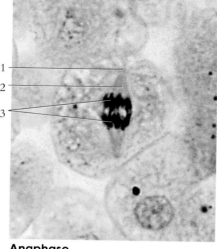

Anaphase
The centromeres split, and the sister chromatids separate as each is pulled to an opposite pole.
1. Aster around centriole
2. Spindle fibers
3. Separating chromosomes

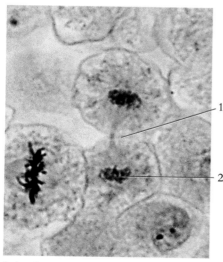

Telophase
The chromosomes lengthen and become less distinct. The cell membrane forms between the forming daughter cells.
1. New cell membrane
2. Newly forming nucleus

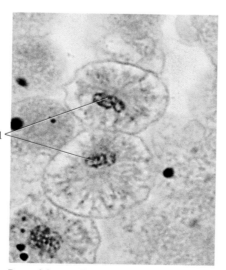

Daughter cells
The single chromosomes (former chromatids—see anaphase) continue to lengthen as the nuclear membrane re-forms. Cell division is complete, and the newly formed cells grow and mature.
1. Daughter nuclei

Figure 2.5 The stages of mitosis followed by cytokinesis. 500X

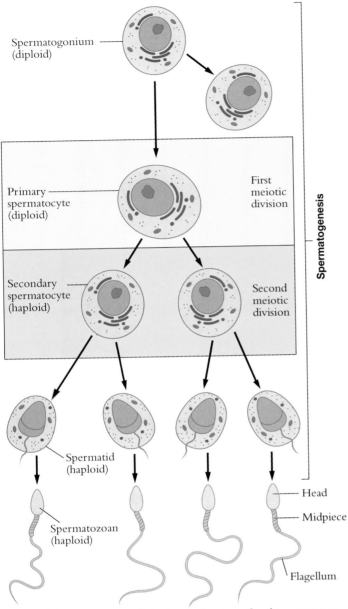

Figure 2.6 Spermatogenesis is the production of male gametes, or spermatozoa, through the process of meiosis.

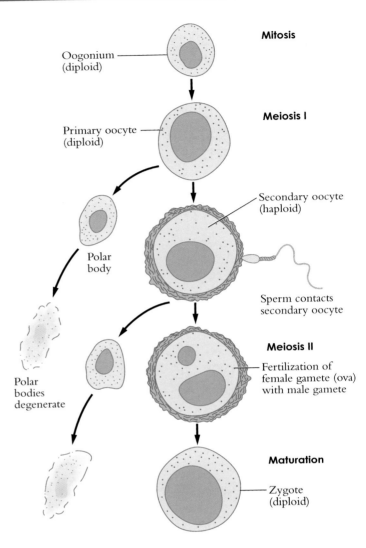

Figure 2.7 Oogenesis is the production of female gametes, or ova, through the process of meiosis.

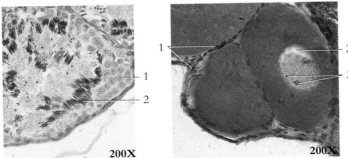

Figure 2.8 Frog testis.
1. Spermatocytes
2. Developing sperm

Figure 2.9 Frog ovary.
1. Follicle cells
2. Germinal vesicle
3. Nucleoli

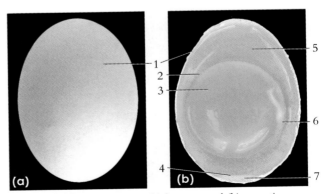

Figure 2.10 (a) An intact chicken egg and (b) a portion of the shell is removed exposing the internal structures.
1. Shell
2. Vitelline membrane
3. Yolk
4. Shell membrane
5. Albumen (egg white)
6. Chalaza (dense albumen)
7. Air space

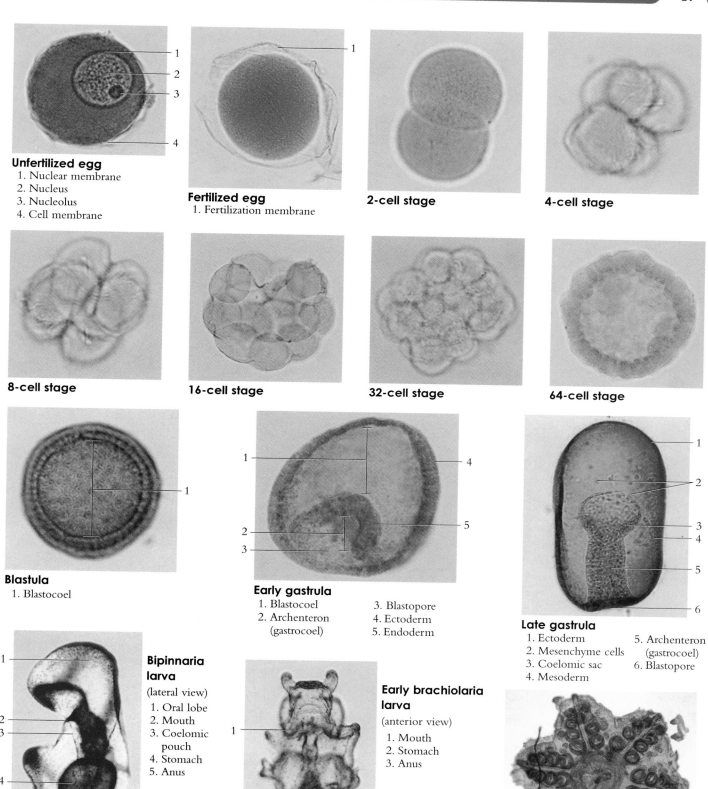

Figure 2.11 Sea star development (all images shown at 100X unless indicated).

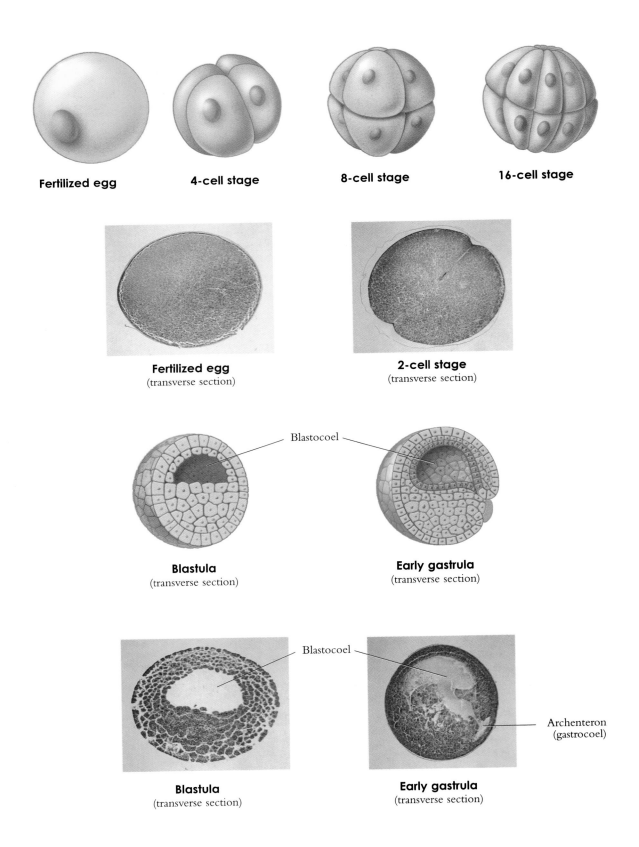

Figure 2.12 Frog development from fertilized egg to early gastrula, shown in diagram and photomicrographs 100X.

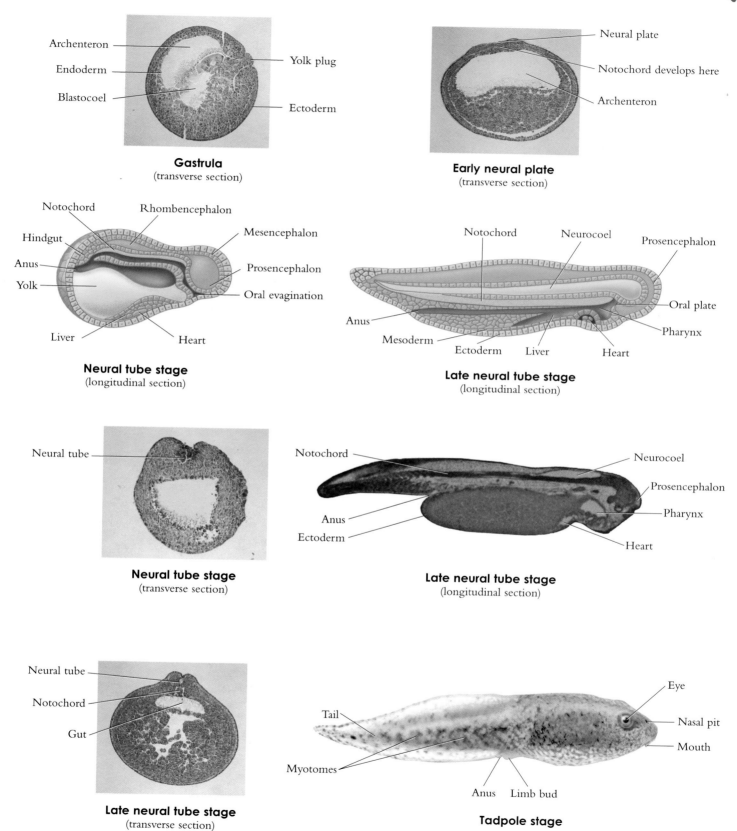

Figure 2.13 Frog development from gastrula to tadpole, shown in diagram and photomicrographs 100X.

A Photographic Atlas for the Zoology Laboratory

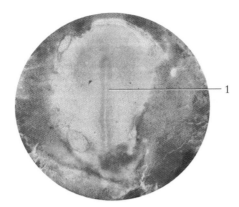

13-hour stage
1. Primitive streak (gastrulation)

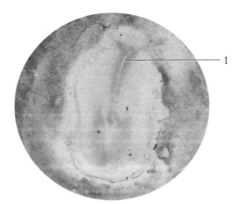

18-hour stage
1. Neurulation beginning

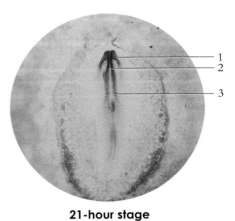

21-hour stage
1. Head fold
2. Neural fold
3. Muscle plate (somites)

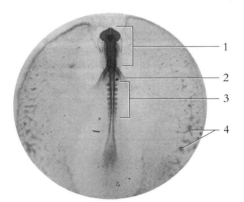

28-hour stage
1. Head fold and brain
2. Artery formation
3. Muscle plate (somites)
4. Blood vessel formation

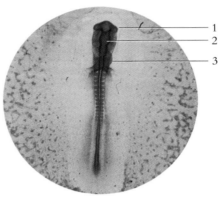

38-hour stage
1. Optic vesicle
2. Brain with five regions
3. Heart

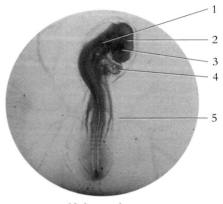

48-hour stage
1. Ear
2. Brain
3. Eye
4. Heart
5. Artery

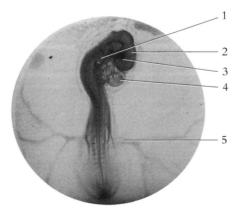

56-hour stage
1. Ear
2. Brain
3. Eye
4. Heart
5. Artery

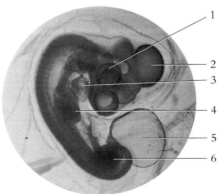

96-hour stage
1. Eye
2. Brain
3. Heart
4. Wing formation
5. Fecal sac
6. Leg formation

Figure 2.14 The major stages of chick development. 20X

Select Single-Celled Eukaryote Supergroup Phyla ("Protists")

Chapter 3

All animals are eukaryotes—their cells contain a membrane-bound nucleus that contains their genetic material. Most eukaryotic cells also contain membrane-bound organelles, such as mitochondria, chloroplasts, and digestive vacuoles and are even capable of meiosis and sexual reproduction. Eukaryotes are most closely related to Archaea, but acquired their organelles from Bacteria by way of endosymbiosis (see exhibit 1 on page vi).

We easily recognize the majority of multicellular animals—the Metazoa—and distinguish these from plants and fungi. But there is a tremendous diversity of eukaryotes that aren't metazoans, fungi, or plants. Some contain chloroplasts, some don't. Most are single-celled, but some aren't. Most are microscopic, but some, like giant kelp, are very large. These organisms, which do not constitute a natural, or monophyletic group, are defined more by what they aren't than by what they are. But because they play an important role in understanding the evolutionary transitions that took place between prokaryotes and metazoans more than a billion years ago, they are crucial components of any serious study of zoology.

Historically, the Linnean classification system ranked taxa according to morphological similarity. As phylogenetic analyses have become increasingly sophisticated and accurate, some of the well-known Linnean taxa have turned out to be evolutionary grades (as opposed to clades), united by primitive (plesiomorphic), as opposed to derived (apomorphic) characters. Such is the case for many independent evolutionary lineages of eukaryotes that are either unicellular or multicellular, but without specialized tissues. Heretofore known as "protists," in this chapter we present them in a phylogenetic context that more accurately reflects their evolutionary history and current taxonomic status.

Most of the unicellular taxa in fig. 3.1 are abundant in aquatic habitats, and many are important constituents of plankton. Plankton are communities of organisms that drift passively or swim slowly in ponds, lakes, and oceans. Plankton are a major source of food for other aquatic organisms. Phototrophic (plantlike) microeukaryotes are major food producers in aquatic ecosystems. Key members of this group are from the Phylum Heterokontophyta, which includes the diatoms and golden algae. The cell wall of a diatom is composed largely of silica rather than cellulose. Some diatoms move in a slow, gliding way as cytoplasm glides through slits in the cell wall.

The Phylum Dinoflagellata also constitutes a large component of the phototrophic plankton. In most species of dinoflagellates, the cell wall is formed of armor-like plates of cellulose. Dinoflagellates are motile, having two flagella. Generally, one encircles the organism in a transverse groove, and the other projects to the posterior.

Among the unicellular microeukaryotes, or 'protozoan' (animallike) phyla are the Amoebozoa, Apicomplexa, Euglenozoa, Metamonada, and Ciliophora. Locomotion of these heterotrophs is by way of flagella, cilia, or pseudopodia of various sorts. In feeding upon other organisms or organic particles, they use simple diffusion, pinocytosis, active transport, or phagocytosis. Although most of these organisms reproduce asexually, some species may also reproduce sexually during a portion of their life cycle. Most protozoa are harmless, although some are parasitic and may cause human disease, including African sleeping sickness and malaria.

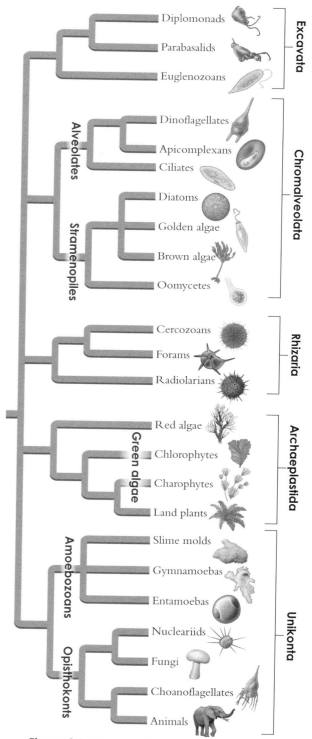

Figure 3.1 Phylogenetic relationships and classification of major eukaryote lineages.

Table 3.1 Representative Single-Celled Eukaryote Supergroup Phyla

Phyla and Representative Kinds	Characteristics
Heterokontophyta — diatoms and golden algae (Plant-like)	Diatom cell walls composed of or impregnated with silica, often with two halves; plastids often golden in Chysophyceae due to chlorophyll composition
Dinoflagellata — dinoflagellates (Plant-like)	Two flagella in grooves of wall; brownish-gold plastids
Amoebozoa — amoebozoa (Animal-like)	Cytoskeleton of microtubules and microfilaments; amoeboid locomotion
Apicomplexa — sporozoa and *Plasmodium* (Animal-like)	Lack locomotor capabilities and contractile vacuoles; mostly parasitic
Euglenozoa — flagellated protozoa (Animal-like)	Use flagella or pseudopodia for locomotion; mostly parasitic
Metamonada — trichomonadas (Animal-like)	Flagellate protozoan, *Trichomonas* sp.
Ciliophora — ciliates and *Paramecium* (Animal-like)	Use cilia to move and feed

Phylum Heterokontophyta - diatoms and golden algae

Figure 3.2 *Biddulphia* sp., a colony-forming diatom. These cells are beginning cell division.

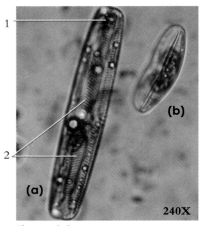

Figure 3.3 Live specimens of pennate (bilaterally symmetrical) diatoms.
(a) *Navicula* sp., and (b) *Cymbella* sp.
1. Chloroplast   2. Striae

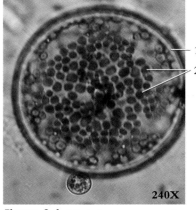

Figure 3.4 *Hyalodiscus* sp., a centric (radially symmetrical) diatom, from a freshwater spring.
1. Silica cell wall   2. Chloroplasts

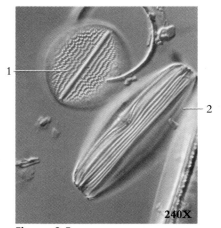

Figure 3.5 Two common freshwater diatoms.
1. *Cocconeis* sp.   2. *Amphora* sp.

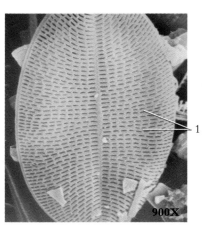

Figure 3.6 A scanning electron micrograph of *Cocconeis* sp., a common freshwater diatom.
1. Striae containing pores, or punctae, in the frustule (silica cell wall).

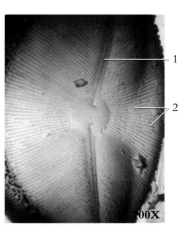

Figure 3.7 A scanning electron micrograph of the diatom *Achnanthes flexella*.
1. Raphe   2. Striae

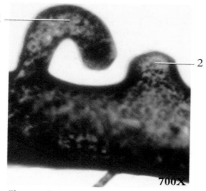

Figure 3.8 A filament with immature gametangia of the "water felt" alga, *Vaucheria* sp. *Vaucheria* is a chrysophyte that is widespread in freshwater and marine habitats. It is also found in the mud of brackish areas that periodically become submerged and then exposed to air.
1. Antheridium
2. Developing oogonium

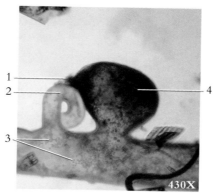

Figure 3.9 *Vaucheria* sp., with mature gametangia.
1. Fertilization pore
2. Antheridium
3. Chloroplasts
4. Developing oogonium

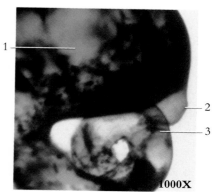

Figure 3.10 *Vaucheria* sp., with mature gametangia.
1. Oogonium
2. Fertilization pore
3. Antheridium

Phylum Dinoflagellata - dinoflagellates

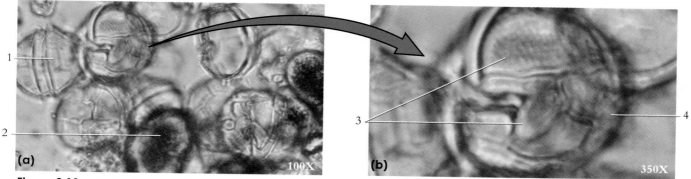

Figure 3.11 Dinoflagellates, *Peridinium* sp. (a) Some organisms are living; (b) others are dead and have lost their cytoplasm and consist of resistant cell walls.
1. Dead dinoflagellate
2. Living dinoflagellate
3. Cellulose plates
4. Remnant of cytoplasm

Figure 3.12 A giant clam with bluish coloration due to endosymbiont dinoflagellates.

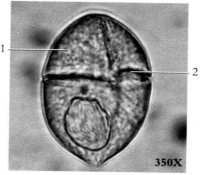

Figure 3.13 A photomicrograph of *Peridinium* sp. The cell wall of many dinoflagellates is composed of overlapping plates of cellulose.
1. Wall of cellulose plates
2. Transverse groove

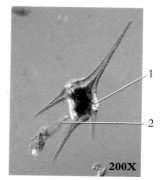

Figure 3.14 *Ceratium* sp. is a common freshwater dinoflagellate.
1. Transverse groove
2. Trailing flagellum

Phylum Amoebozoa - amoebas

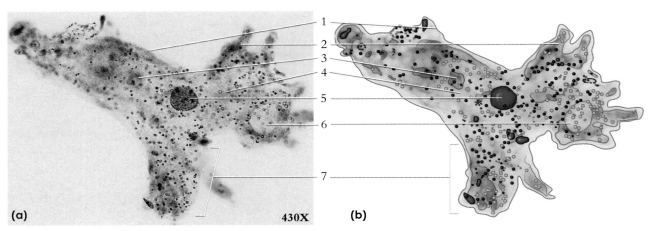

Figure 3.15 The *Amoeba proteus* is a fresh-water protozoan that moves by forming cytoplasmic extensions called pseudopodia. (a) Stained cell, and (b) diagram.
1. Cell membrane
2. Ectoplasm
3. Food vacuole
4. Endoplasm
5. Nucleus
6. Contractile vacuole
7. Pseudopodia

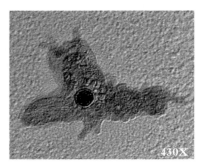

Figure 3.16 *Amoeba proteus* (stained blue).

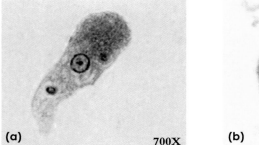

Figure 3.17 Protozoan *Entamoeba histolytica* is the causative agent of amebic dysentery, a disease most common in areas with poor sanitation. (a) A trophozoite, and (b) a cyst.

Phylum Apicomplexa - *Plasmodium*

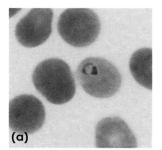 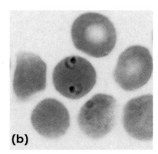 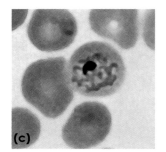 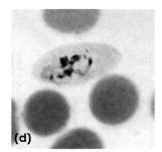

Figure 3.18 The protozoan *Plasmodium falciparum* causes malaria, which is transmitted by the female *Anopheles* mosquito. (a) The ring stage in a red blood cell, (b) a double infection, (c) a developing schizont, and (d) a gametocyte.

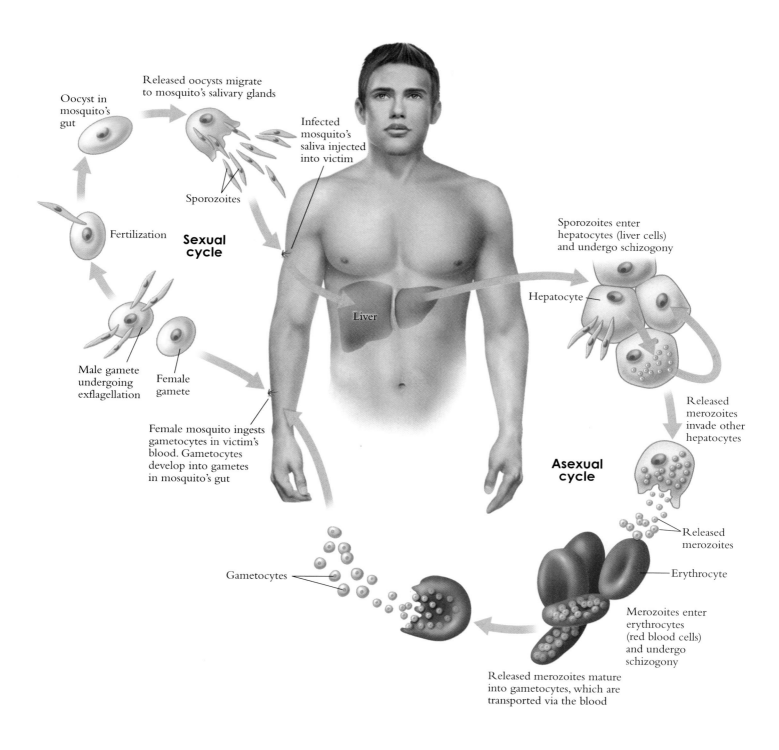

Figure 3.19 The life cycle of the protozoan *Plasmodium vivax* that is carried by the female *Anopheles* mosquito, which causes malaria in humans. During the sexual cycle, sporozoites are produced in the mosquito. During the asexual cycle, merozoites are first produced in hepatocytes and then in erythrocytes.

Phylum Metamonada - (Trichomonas) and Phylum Euglenozoa - (Leishmania and Trypanosoma): flagellated protozoans

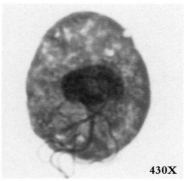

Figure 3.20 The protozoan *Trichomonas vaginalis* is the causative agent of trichomoniasis. Trichomoniasis is an inflammation of the genitourinary mucosal surfaces—the urethra, vulva, vagina, and cervix in females and the urethra, prostate, and seminal vesicles in males.

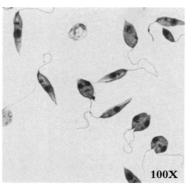

Figure 3.21 The protozoan *Leishmania donovani* is the causative agent of leishmaniasis, or kala-azar disease, in humans. The sandfly is the infectious host of this disease.

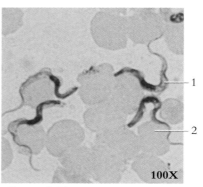

Figure 3.22 A flagellated protozoan *Trypanosoma brucei* is the causative agent of trypanosomiasis, or African sleeping sickness. The tsetse fly is the infectious host of this disease in humans.
1. *Trypanosoma brucei*
2. Red blood cell

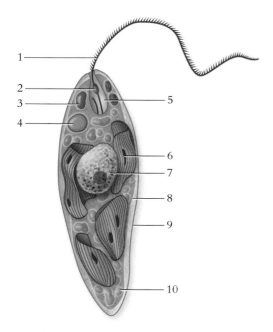

Figure 3.23 A diagram of *Euglena* sp., a genus of flagellates that contain chloroplasts. They are freshwater organisms that have a flexible pellicle rather than a rigid cell wall.

1. Long flagellum
2. Photoreceptor
3. Eye spot
4. Contractile vacuole
5. Reservoir
6. Chloroplast
7. Nucleus
8. Pellicle
9. Cell membrane
10. Paramylon granule

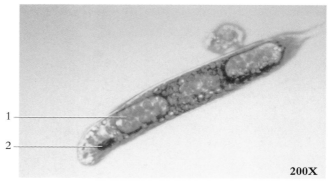

Figure 3.24 A species of *Euglena*.
1. Paramylum body
2. Photoreceptor

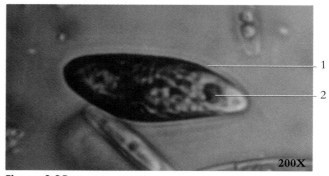

Figure 3.25 A species of *Euglena* from a brackish lake.
1. Pellicle
2. Photoreceptor

Select Single-Celled Eukaryote Supergroup Phyla ("Protists")

Phylum Ciliophora - ciliates and paramecia

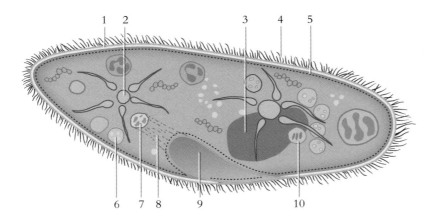

Figure 3.26 *Paramecium caudatum* is a ciliated protozoan. The poisonous trichocysts of these unicellular organisms are used for defense and capturing prey.
1. Pellicle
2. Contractile vacuole
3. Macronucleus
4. Cilia
5. Trichocyst
6. Food vacuole
7. Forming food vacuole
8. Gullet
9. Oral groove
10. Micronucleus

Figure 3.27 *Paramecium caudatum* is a ciliated protozoan. Paramecia are usually common in ponds containing decaying organic matter.
1. Macronucleus
2. Contractile vacuole
3. Micronucleus
4. Pellicle
5. Cilia

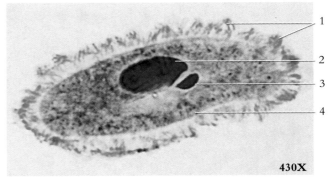

Figure 3.28 *Paramecium bursaria* is a unicellular, slipper-shaped organism. When disturbed or threatened, they release spear-like trichocysts as a defense.
1. Trichocysts
2. Macronucleus
3. Micronucleus
4. Pellicle

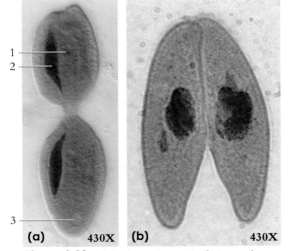

Figure 3.29 (a) A *Paramecium* sp. in fission and (b) a *Paramecium* sp. in conjugation.
1. Micronucleus
2. Macronucleus
3. Contractile vacuole

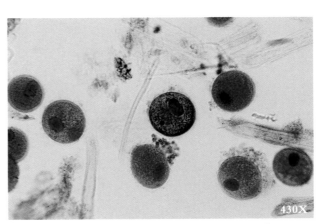

Figure 3.30 *Balantidium coli* is the causative agent of balantidiasis. Cysts in sewage-contaminated water are the infective form.

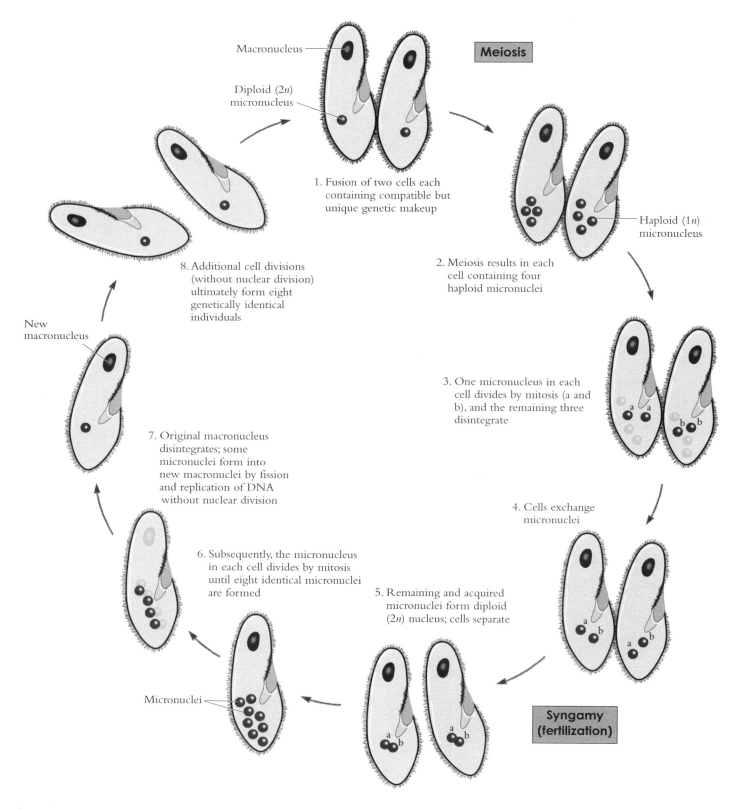

Figure 3.31 The sexual shuffling of genes during conjugation and genetic recombination in *Paramecium*.

Porifera

Chapter 4

An estimated 8,340 species of sponges are contained within the phylum *Porifera*. Sponges are mostly marine organisms that lack differentiated tissues and body symmetry. The body of a sponge consists of masses of cells embedded in a supporting gelatinous matrix called mesohyl. The body is perforated by many pores for the passage of water.

Adult sponges are sessile, or anchored in place. Adult sponges obtain food particles through the forced circulation of water through their bodies. Water enters the central cavity, or *spongocoel*, through numerous pores, called *ostia*, and flows out of the body through the *osculum*. Water is kept moving by the action of flagellated *choanocytes*, or *collar cells*. Choanocytes obtain food particles from the water by phagocytosis. Wandering cells, called *amoebocytes*, transport nutrients from the choanocytes to other body cells. The body of a sponge is structurally supported by calcium carbonate or silica projections, called *spicules*, and by fibers of a tough protein called *spongin*.

Sponges reproduce sexually as eggs and sperm are released into the water, where fertilization occurs. The zygote develops into a free-swimming larva that soon attaches and matures into a sponge.

Sponges are a source of food for many marine animals. Sponges are harvested and prepared for commercial use by beating them to soften the spicule and spongin supporting structures, and then drying them in the sun. The soft, pliable, and absorbent nature of a prepared sponge carcass makes it ideal for wiping and cleaning.

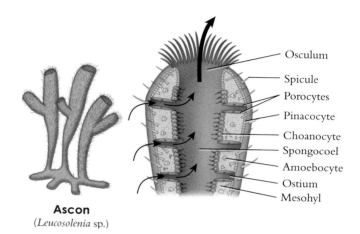

Ascon
(*Leucosolenia* sp.)

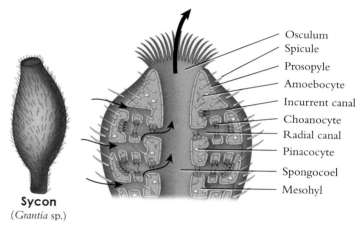

Sycon
(*Grantia* sp.)

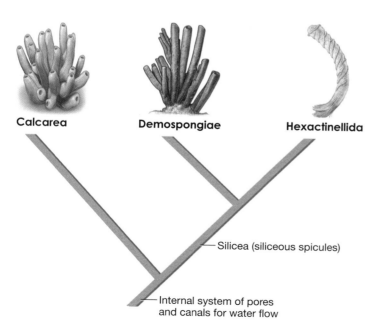

Figure 4.1 Phylogenetic relationships and classification of Porifera.

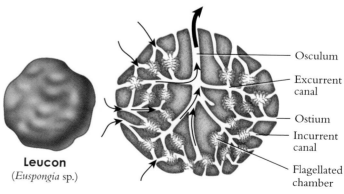

Leucon
(*Euspongia* sp.)

Figure 4.2 Examples of sponge body types. A diagrammatic representative of each of the three types depicts with arrows the flow of water through the body of the sponge.

Class Calcarea

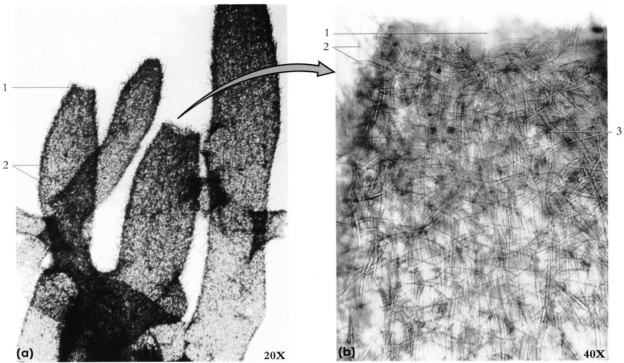

Figure 4.3 (a) *Leucosolenia* sp. has an ascon body type. (b) A high magnification of the spicules and ostia.
1. Osculum 2. Spicules 3. Ostia

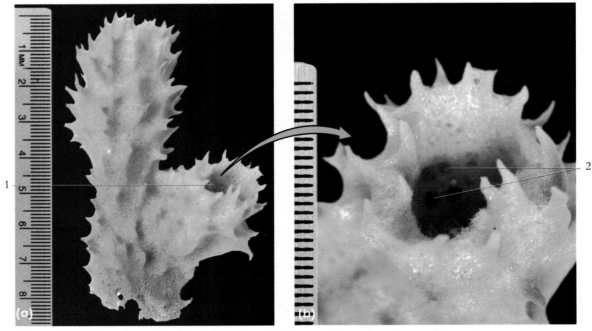

Figure 4.4 (a) A sponge with an ascon body type. (b) Close-up view of osculum (scale in mm).
1. Osculum 2. Ostia (seen from inside osculum)

Porifera

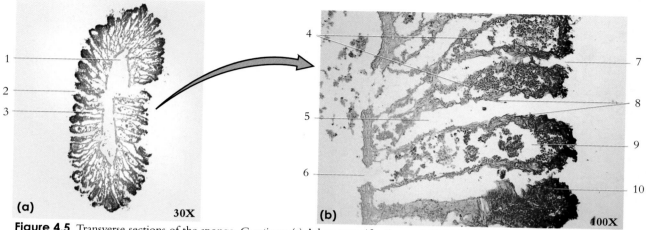

Figure 4.5 Transverse sections of the sponge, *Grantia* sp. (a) A low magnification and (b) high magnification.
1. Spongocoel
2. Ostium (incurrent canal)
3. Radial canal
4. Choanocytes (collar cells)
5. Incurrent canal
6. Apopyle
7. Ostium
8. Pinacocytes
9. Radial canal
10. Mesohyl

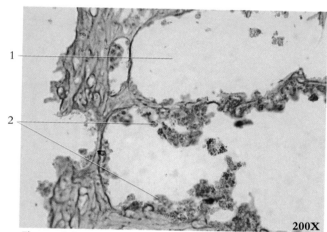

Figure 4.6 A transverse section of the sponge, *Grantia* sp., showing collar cells.
1. Radial canal
2. Choanocytes (collar cells)

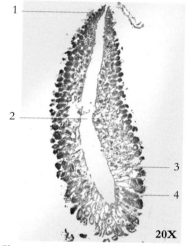

Figure 4.7 A longitudinal section of the sponge, *Grantia* sp.
1. Osculum
2. Spongocoel
3. Radial canal
4. Ostium (incurrent canal)

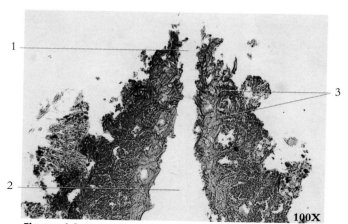

Figure 4.8 A longitudinal section of the sponge, *Grantia* sp., showing magnified view of osculum.
1. Osculum
2. Spongocoel
3. Spicules

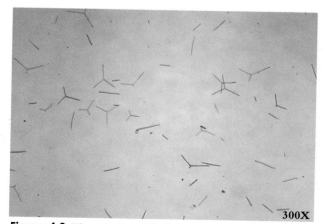

Figure 4.9 The spicules of *Scypha* sp.

Class Demospongiae

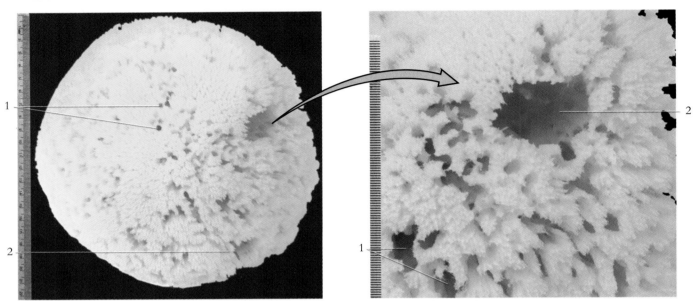

Figure 4.10 A bath sponge, class Demospongiae, has a leuconoid body structure (scale in mm).
1. Ostia
2. Osculum

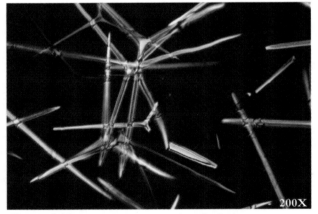

Figure 4.11 Branched silica spicules of a freshwater sponge.

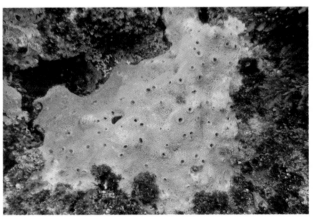

Figure 4.12 An encrusting sponge. Leuconoid sponges display a wide range of color and shape.

Figure 4.13 A leuconoid sponge growing next to a tunicate (scale in mm).
1. Sponge
2. Tunicate

Figure 4.14 A yellow ball sponge, *Cinachyra allocladia* (scale in mm).

Ctenophora and Cnidaria

Chapter 5

There are approximately 240 species within the phylum *Ctenophora*. They are commonly known as comb jellies. The phylogenetic position of ctenophores relative to the other metazoans is tenuous, but very important because of their potential to provide insights on some of the earliest aspects of animal evolution. Currently, the most strongly supported hypothesis supports the idea that ctenophores are sister taxon to the rest of the metazoans.

Most ctenophores are hermaphroditic. Development of the fertilized eggs is direct, with no distinctive larval form. Juveniles lack tentacles and tentacle sheaths, and in most species the juveniles gradually develop the body forms of their parents. They have eight rows of fused cilia used for locomotion. Body types are quite diverse within the phylum, especially considering the small number of species.

Almost all ctenophores are predators preying primarily on zooplankton. They employ a wide variety of techniques to capture their prey. Ctenophores lack stinging cells. In order to capture prey ctenophores rely on sticky cells called colloblasts. In a few species, special cilia in the mouth are used for biting gelatinous prey.

There are an estimated 10,100 species within the phylum *Cnidaria*. It is a large and diverse group of simply organized aquatic (mostly marine) animals that includes hydras, jellyfishes, sea anemones, and colonial corals. There are two morphological types of cnidarians: the *polyp*, or hydroid form, and the *medusa*, or jellyfish form. The polyp forms are usually sessile, or anchored in place. The medusa forms are floating or free-swimming. Most cnidarians exhibit both body types at some point in their life cycle.

Adult cnidarians have a *radially symmetrical body* that is a simple sac with the mouth opening into the *coelenteron (gastrovascular cavity)*. Food wastes are also eliminated through the mouth. The mouth is usually surrounded by tentacles bearing stinging cells called *cnidocytes*. The body is composed of an outer *epidermis* and an inner gastrodermis. A gelatinlike *mesoglea* separates these two layers.

Polyp types of cnidarians have asexual reproduction by budding. Medusa types and a few polyp types of cnidarians have sexual reproduction by gametes.

As carnivores, cnidarians are important predators within the marine food chain. They obtain their food by using their cnidocytes to sting and paralyze their prey. In tropical waters, the corals of the class Anthozoa form massive colonies. Their skeletons remain in place even after the polyps die and form the basis of coral reefs and some oceanic islands.

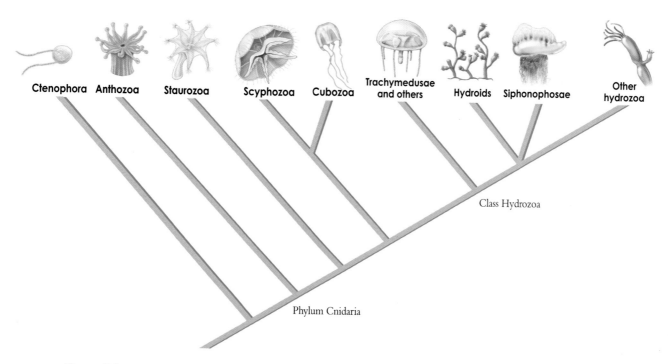

Figure 5.1 Phylogenetic relationships and classification of Ctenophora and Cnidaria.

Table 5.1 Representatives of the Phylum Ctenophora

Classes and Representative Kinds	Characteristics
Tentaculata — comb jellies	Marine coastal waters; utilize cilia for transportation; most species hermaphroditic; lack stinging cells

Representatives of the Phylum Cnidaria

Classes and Representative Kinds	Characteristics
Hydrozoa — hydra, *Obelia*, and Portuguese man-of-war	Mainly marine; both polyp and medusa stage (polyp form only in hydra); polyp colonies in most
Scyphozoa — jellyfish	Marine coastal waters; polyp stage restricted to small larval forms
Cubozoa — box jellyfish	Marine coastal waters; polyp and medusa stage; square-shaped when viewed from above
Anthozoa — sea anemones, corals, and sea fans	Marine coastal waters; solitary or colonial polyps; no medusa stage; partitioned gastrovascular cavity

Phylum Ctenophora

Class Tentaculata

Figure 5.2 The warty comb jelly, *Mnemiopsis leidyi*, is commonly found in the western Atlantic.
1. Rows of cilia

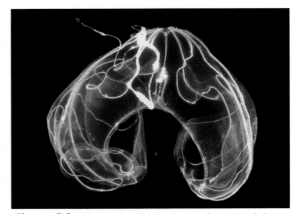

Figure 5.3 The comb jelly, *Bathocyroe fosteri*, is a lobate species found in all the world's oceans.

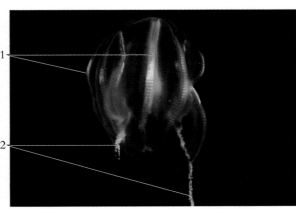

Figure 5.4 The Arctic comb jelly or sea nut, *Mertensia ovum*, is found in polar seas.
1. Rows of cilia   2. Tentacles (left tenticle is retracted)

Figure 5.5 *Beroe cucumis* is found in the seas surrounding the United Kingdom.

Phylum Cnidaria
Class Hydrozoa

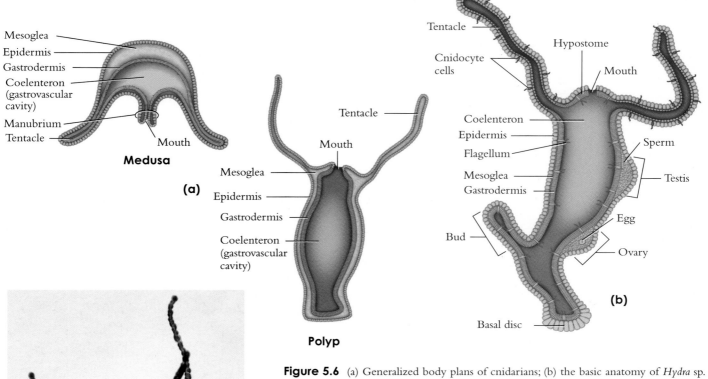

Figure 5.6 (a) Generalized body plans of cnidarians; (b) the basic anatomy of *Hydra* sp.

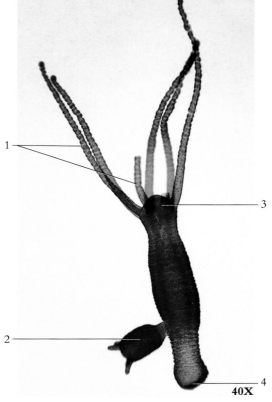

Figure 5.7 A budding *Hydra* sp.
1. Tentacles
2. Bud
3. Hypostome
4. Basal disc (foot)

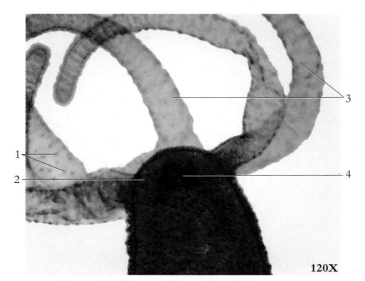

Figure 5.8 The anterior end of a *Hydra* sp.
1. Cnidocytes
2. Hypostome
3. Tentacles
4. Mouth

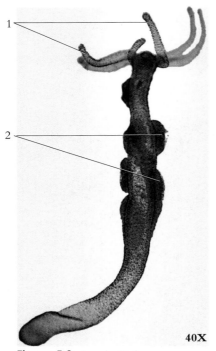

Figure 5.9 A male *Hydra* sp.
1. Tentacles
2. Testes

Figure 5.10 A female *Hydra* sp.
1. Tentacles
2. Ovary
3. Basal disc (foot)

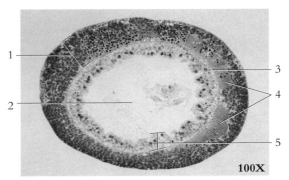

Figure 5.11 A transverse section of a female *Hydra* sp.
1. Epidermis (ectoderm)
2. Coelenteron
3. Mesoglea
4. Eggs
5. Gastrodermis (endoderm)

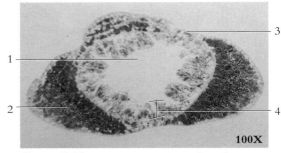

Figure 5.12 A transverse section of a male *Hydra* sp.
1. Coelenteron
2. Testes
3. Epidermis (ectoderm)
4. Gastrodermis (endoderm)

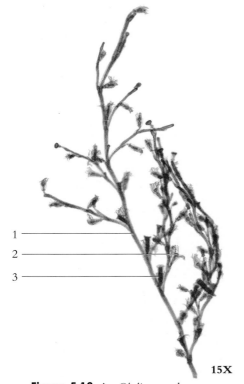

Figure 5.13 An *Obelia* sp. colony.
1. Coenosarc (soft tissue connecting polyps)
2. Hydranth (feeding polyp)
3. Gonangium (reproductive polyp)

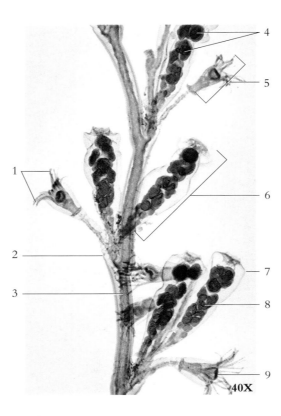

Figure 5.14 A detail of an *Obelia* sp. colony.
1. Tentacles
2. Perisarc (horny covering that encloses the polyp)
3. Coenosarc
4. Medusa buds
5. Hydranth (feeding polyp)
6. Gonangium (reproductive polyp)
7. Gonotheca (protective covering)
8. Blastostyle
9. Hypostome

Ctenophora and Cnidaria

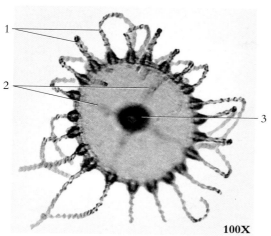

Figure 5.15 An aboral view of an *Obelia* sp. medusa.
1. Tentacles
2. Radial canals
3. Manubrium (seen through the body from above)

Figure 5.16 *Obelia* sp. medusa in feeding position.
1. Tentacles
2. Gonad
3. Manubrium
4. Mouth

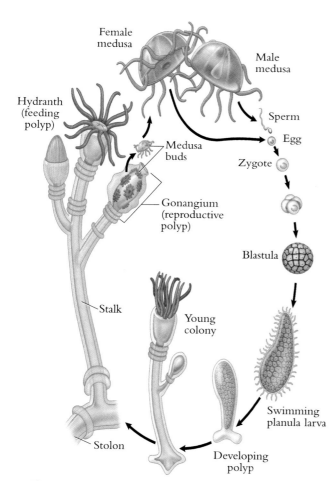

Figure 5.17 The life cycle of *Obelia* sp.

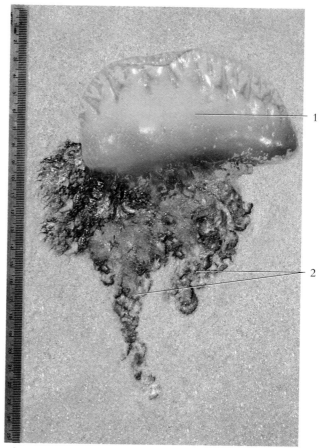

Figure 5.18 The Portuguese man-of-war, *Physalia* sp., is actually a colony of medusae and polyps acting as a single organism. The tentacles are composed of three types of polyps: the gastrozooids (feeding polyps), the dactylozooids (stinging polyps), and the gonozooids (reproductive polyps) (scale in mm).
1. Pneumatophore (float)
2. Tentacles

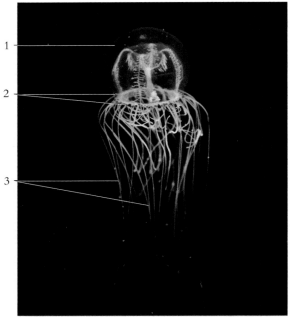

Figure 5.19 The red-eye jellyfish, *Polyorchis penicillatus,* ranges from the Aleutian Islands in Alaska to Sea of Cortez in Mexico. It is a common bell jelly found near the shoreline.
1. Bell
2. Ocelli (eyes)
3. Tentacles

Figure 5.20 The blue button jelly, *Porpita porpita*, consists of two main parts, a colony of hydroids and a float. It is found in tropical waters.
1. Float
2. Hydroid colony (bluish-green color)

Class Scyphozoa

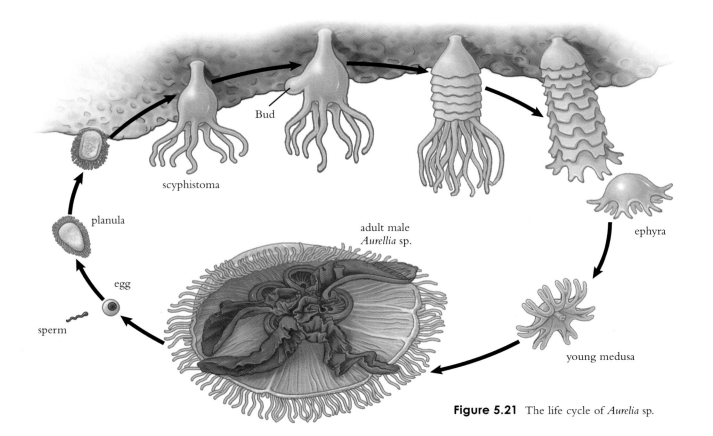

Figure 5.21 The life cycle of *Aurelia* sp.

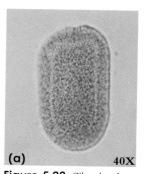 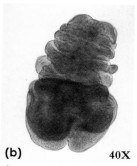 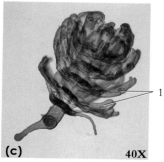 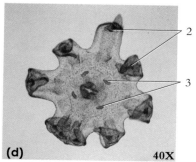

Figure 5.22 The developmental stages of *Aurelia* sp. (a) Planula larvae develops from a fertilized egg that may be retained on the oral arm of the medusa. The scyphistoma (b) develops into the strobila (c) from which the ephyra larvae (d) break free and develop into adult jellyfish.
1. Developing ephyrae
2. Rhopalia (sense organs)
3. Gonads

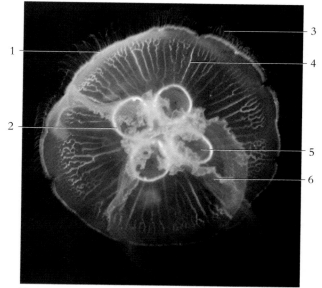

Figure 5.23 An oral view of *Aurelia* sp. medusa.
1. Ring canal
2. Gonad
3. Marginal tentacles
4. Radial canal
5. Subgenital pit
6. Oral arm

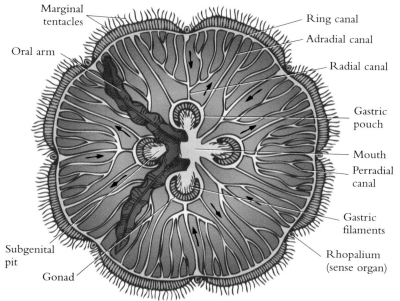

Figure 5.24 An oral view of *Aurelia* sp. medusa. The right oral arms have been removed, and the arrows depict circulation through the canal system.

Figure 5.25 The red-striped jellyfish, *Chrysaora melanasteris*, is common near the surface of the Bering Sea.

Figure 5.26 The purple-striped jelly, *Chrysaora colorata*, is found off the coast of California and in Monterey Bay.

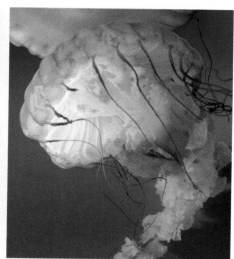

Figure 5.27 The sea nettle, *Chrysaora fuscescens*, often masses in large swarms off the Pacific coast.

Class Cubozoa

Figure 5.28 The box jellyfish, *Carybdea sivickisi*, is named for its cube-shaped bell. All cubozoans have four tentacles.

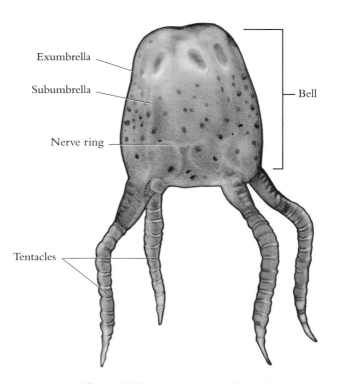

Figure 5.29 An illustration of box jellyfish, *Carybdea sivickisi*, showing basic external structures.

Class Anthozoa

Figure 5.30 The sunburst anemone, *Anthopleura sola*, gets its green coloration from symbiotic algae within it.

Figure 5.31 The firecracker coral, *Dendrophyllia* sp., a filter feeder, actively feeds day and night.

Figure 5.32 The tube anemone, *Pachycerianthus fimbriatus*, makes a leathery tube and sinks it up to two feet into the sand.

Figure 5.33 The sea pen, *Ptilosarcus gurneyi*, is a colony of polyps that may reach two feet in height.

Figure 5.34 Disk anemones, *Actinodiscus* sp., form large colonies.

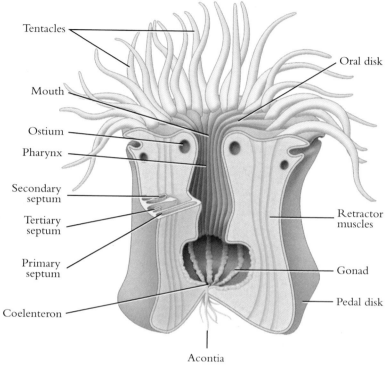

Figure 5.35 A diagram of a partially dissected sea anemone, *Metridium* sp.

Figure 5.36 A group of anemones, *Anthopleura* sp., in a tide pool with other tidal organisms.

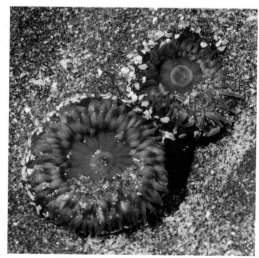

Figure 5.37 Sunburst anemones, *Anthopleura sola*, at low tide.

Figure 5.38 Brain coral, *Goniastrea* sp.

Figure 5.39 The skeletal structure of brain coral, *Goniastrea* sp.

Figure 5.40 Mushroom coral, *Rhodactis* sp.

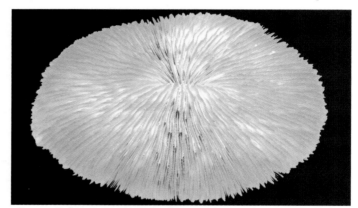
Figure 5.41 The skeletal structure of mushroom coral, *Rhodactis* sp.

Figure 5.42 Elkshorn coral, *Acropora* sp.

Figure 5.43 The skeletal structure of elkshorn coral, *Acropora* sp.

Figure 5.44 A detailed view of the polyps of candy cane coral, *Caulastrea furcata*.

Figure 5.45 A detailed view of the polyps of glove xenia, *Xenia umbellata*.

Platyhelminthes

Chapter 6

An estimated 29,280 species of flatworms are contained within the phylum *Platyhelminthes*. Flatworms are aquatic or parasitic in the body of a host animal. They are soft-bodied, flattened animals that include planaria, flukes, and tapeworms. The *bilaterally symmetrical body* of a flatworm is composed of three tissue layers (ectoderm, mesoderm, and endoderm) and has a distinct head with a simple brain consisting of two masses of nervous tissue called *ganglia*. *Nerve cords* from the ganglia extend the length of the body. Excretory organs, called *protonephridia*, consist of *flame cells* in the body tissues and branched tubules that extend through the body and exit through pores at the body surface. The mouth opens into the *gastrovascular cavity*. The reproductive organs are well developed.

Because the human is a host animal for many flatworms, these parasitic animals are of major health concern. In tropical countries they cause high numbers of deaths, especially in children. Parasitic flatworms ingest significant quantities of nutrients, secrete toxic wastes, and generally interfere with normal physiological processes.

Table 6.1 Some Representatives of the Phylum Platyhelminthes

Classes and Representative Kinds	Characteristics
Turbellaria — planarians	Mostly free-living, carnivorous, aquatic forms; body covered by ciliated epidermis
Trematoda — flukes including schistosomes	Parasitic with wide range of invertebrate and vertebrate hosts; suckers for attachment to host
Cestoda — tapeworms	Parasitic in many vertebrate hosts; complex life cycle with intermediate hosts; suckers or hooks on scolex for attachment to host; eggs are produced and shed within proglottids

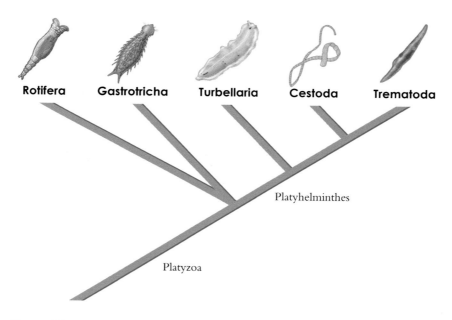

Figure 6.1 Phylogenetic relationships and classification of representative flatworms (class Monogenea is not depicted but forms a monophyletic group with Cestoda and Trematoda; Turbellaria is likely a paraphyletic group).

Class Turbellaria

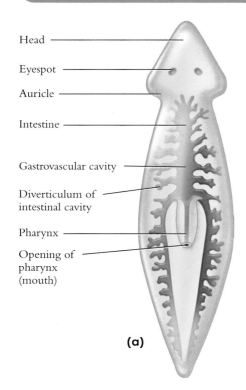

(a)

- Head
- Eyespot
- Auricle
- Intestine
- Gastrovascular cavity
- Diverticulum of intestinal cavity
- Pharynx
- Opening of pharynx (mouth)

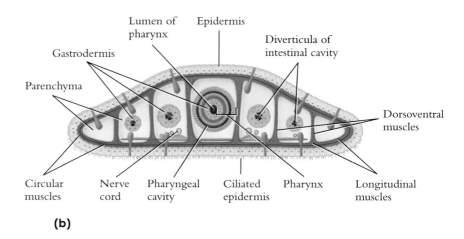

(b)

- Lumen of pharynx
- Epidermis
- Gastrodermis
- Diverticula of intestinal cavity
- Parenchyma
- Dorsoventral muscles
- Circular muscles
- Nerve cord
- Pharyngeal cavity
- Ciliated epidermis
- Pharynx
- Longitudinal muscles

Figure 6.2 The internal anatomy of planarian. (a) A longitudinal section, and (b) a transverse section through the pharyngeal region.

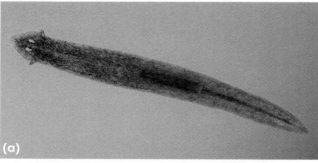
(a)

(b)

Figure 6.3 A planarian (a) *Dugesia* sp. is aquatic, while (b) *Bipalium* sp. is a common inhabitant of gardens.

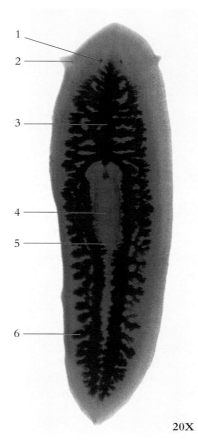

20X

Figure 6.4 *Dugesia* sp.
1. Eyespot
2. Auricle
3. Gastrovascular cavity
4. Pharynx
5. Opening of pharynx (mouth)
6. Diverticulum of intestinal cavity

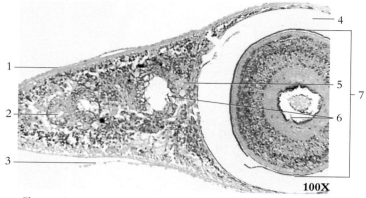

Figure 6.5 A transverse section through the pharyngeal region of *Dugesia* sp.
1. Epidermis
2. Testis
3. Cilia
4. Pharyngeal cavity
5. Dorsoventral muscles
6. Gastrodermis (endoderm)
7. Pharynx

Figure 6.6 A transverse section through the posterior region of *Dugesia* sp.
1. Epidermis
2. Intestinal cavity
3. Mesenchyme
4. Dorsoventral muscles
5. Gastrodermis (endoderm)

Class Trematoda

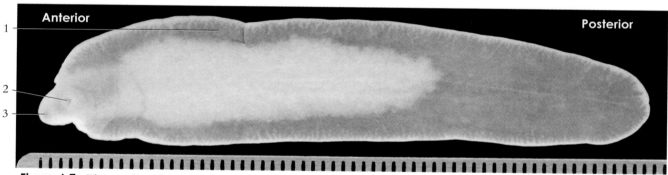

Figure 6.7 The cow liver fluke, *Fasciola magna*, is one of the largest flukes, measuring up to 7.75 cm long (scale in mm).
1. Yolk gland
2. Ventral sucker
3. Oral sucker

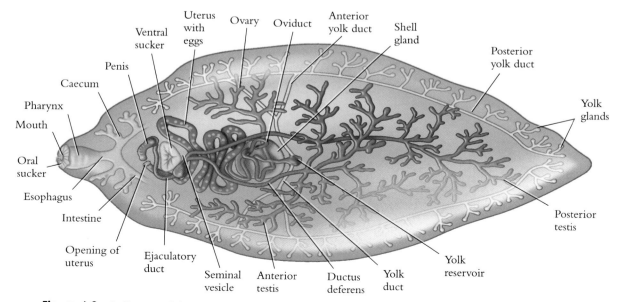

Figure 6.8 A diagram of the sheep liver fluke, *Fasciola hepatica*.

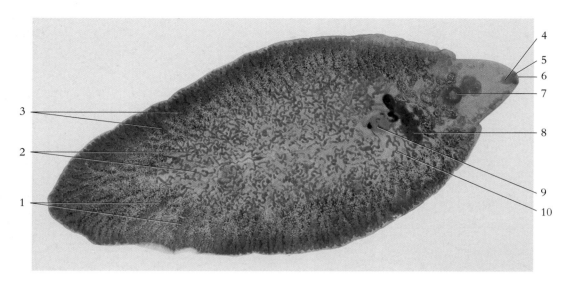

Figure 6.9 The sheep liver fluke, *Fasciola hepatica*.
1. Caecum
2. Testis
3. Yolk duct
4. Pharynx
5. Oral sucker
6. Mouth
7. Ventral sucker
8. Uterus with eggs
9. Shell gland
10. Ovary

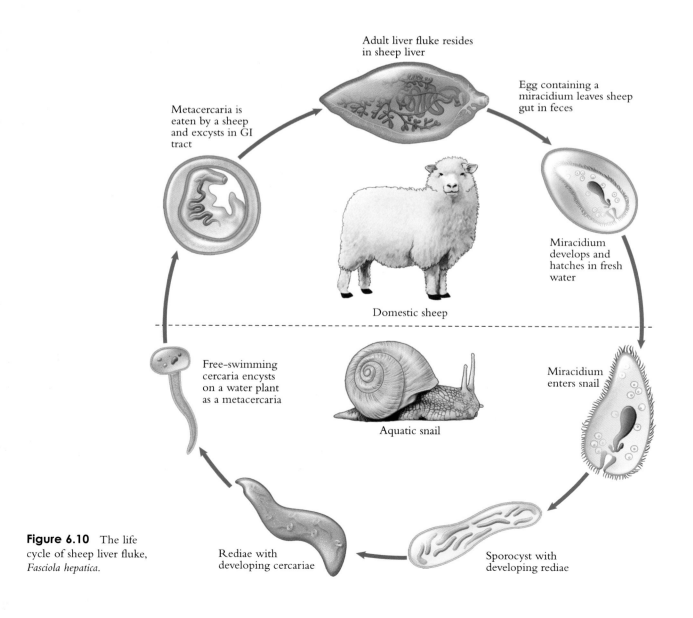

Figure 6.10 The life cycle of sheep liver fluke, *Fasciola hepatica*.

Platyhelminthes 47

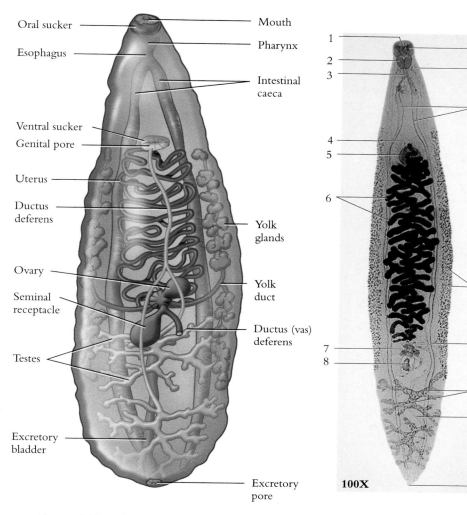

Figure 6.11 A diagram of the human liver fluke, *Clonorchis sinensis*.

Figure 6.12 The liver fluke, *Clonorchis* sp.
1. Mouth
2. Pharynx
3. Esophagus
4. Genital pore
5. Ventral sucker
6. Uterus
7. Ovary
8. Seminal receptacle
9. Oral sucker
10. Cerebral ganglion
11. Intestinal caeca
12. Yolk glands
13. Yolk duct
14. Testis
15. Ductus (vas) deferens
16. Excretory pore

Figure 6.13 The cercaria stage of a trematode species.

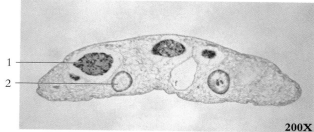

Figure 6.14 A transverse section through the midbody region of *Clonorchis* sp.
1. Uterus
2. Intestine

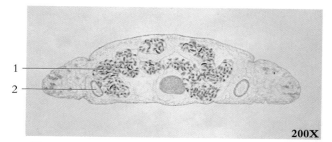

Figure 6.15 A transverse section through the lower body region of *Clonorchis* sp.
1. Testis
2. Intestine

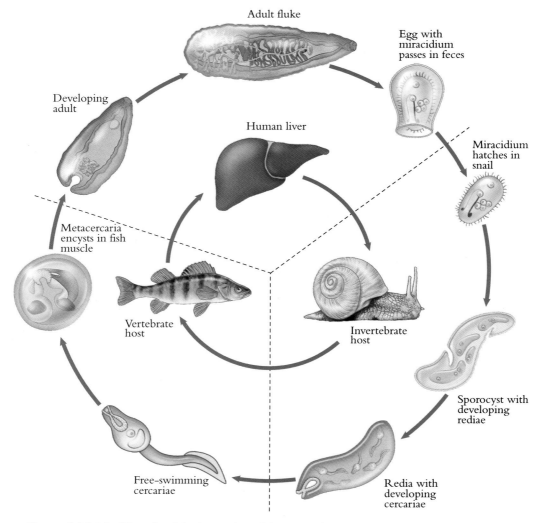

Figure 6.16 The life cycle of the human liver fluke, *Clonorchis sinesis*.

Class Cestoda

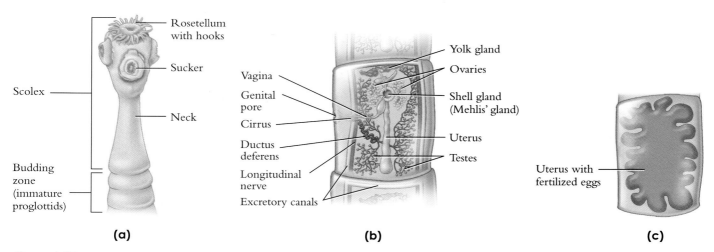

Figure 6.17 The diagrams of a parasitic tapeworm, *Taenia pisiformis*. (a) The anterior end, (b) mature proglottids, and (c) a gravid proglottid.

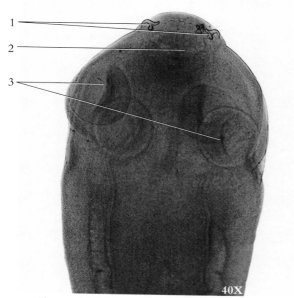

Figure 6.18 The scolex of *Taenia pisiformis*.
1. Hooks
2. Rostellum
3. Suckers

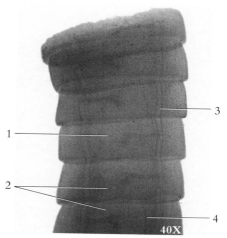

Figure 6.19 The immature proglottids of *Taenia pisiformis*.
1. Early ovary
2. Early testes
3. Excretory canal
4. Immature vagina and ductus deferens

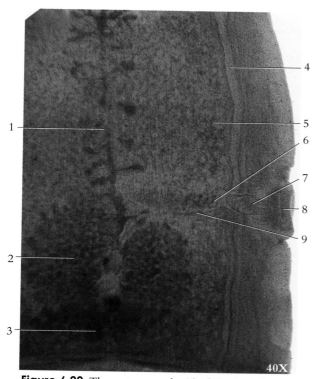

Figure 6.20 The mature proglottid of *Taenia pisiformis*.
1. Uterus
2. Ovary
3. Yolk gland
4. Excretory canal
5. Testes
6. Ductus deferens
7. Cirrus
8. Genital pore
9. Vagina

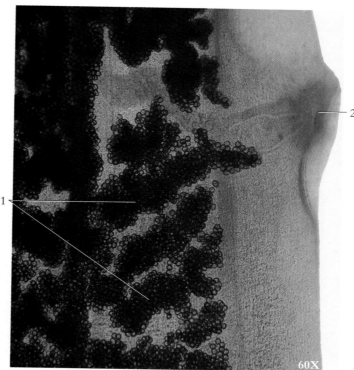

Figure 6.21 The gravid proglottid of *Taenia pisiformis*.
1. Zygotes in branched uterus
2. Genital pore

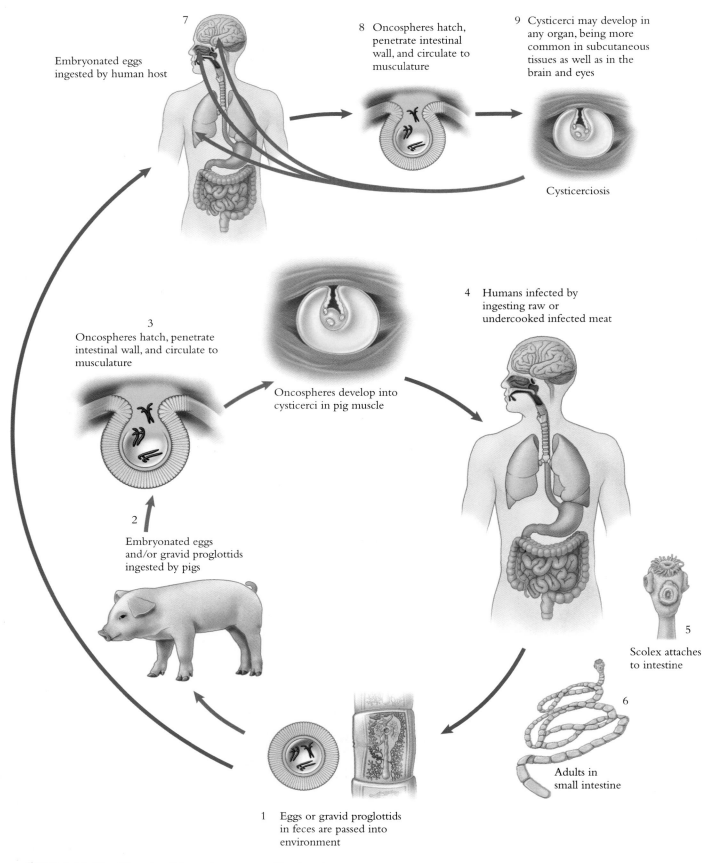

Figure 6.22 The life cycle of the pork tapeworm, *Taenia saginata*.

Mollusca, Brachiopoda, and Bryozoa

Chapter 7

An estimated 117,350 species of mollusks are contained within the phylum *Mollusca*. Mollusks are aquatic and terrestrial soft-bodied or shelled animals that include such forms as snails, clams, oysters, squids, and octopuses. They are *bilaterally symmetrical* and have true *coeloms* and usually distinct heads. Many mollusks have a muscular foot for locomotion and a soft visceral mass enclosed by a heavy fold of tissue called the *mantle*. Many species have a protective *shell*, which is secreted by the mantle.

Mollusca is a large and diverse phylum, with species ranging in size from a small snail that is barely macroscopic to the giant squid that is the largest invertebrate animal. Mollusks are important in marine and freshwater food chains. Some are consumed as food by humans and are of great commercial importance. These include clams, oysters, snails, mussels, squid, and octopuses. Other mollusks are hosts for disease-causing parasites and are of medical importance. These include a number of species of snails. In addition, some species of slugs and snails are of economic importance because they are devastating pests to certain crop plants.

Superficially, brachiopods resemble bivalves, but their hard shells are arranged on the upper and lower halves of the animal, as opposed to the left- and right-hand orientation of bivalves. Another difference is that bivalves use their gills to filter phytoplankton and other nutrients from the water, whereas brachiopods use a specialized organ called a lophophore. Brachiopods also lack the bivalve's siphon and foot, which restricts their options for movement.

Brachiopods are entirely marine, but have a rich fossil record that extends back to the early Cambrian Period. Although they once numbered over 12,000 species, only about 440 remain. Some brachiopods, such as *Ligulata*, are virtually identical to 400-million-year-old fossils. This fossil record serves as an important indicator of changes in climate and biodiversity over the course of Earth's history.

There are over 4,000 extant bryozoan species, most of which are marine, but a large number are found in freshwater. Bryozoans are filter feeders, sharing evolutionary affinities with the Brachiopoda and Phoronida, including their specialized filter feeding organ, the lophophore. The lophophore consists of tentacles lined with cilia. The cilia capture plankton and detritus from the water column, and the tentacles move it to the mouth. Many bryozoan species are polyandrous—that is, they start out as male and later become female. Most are sessile, some of which are persistent fouling organisms on ship hulls, docks, water pipes, and sewage treatment facilities.

Table 7.1 Representatives of the Phylum Mollusca

Classes and Representative Kinds	Characteristics
Polyplacophora — chitons	Marine; shell of eight dorsal plates; broad foot
Gastropoda — snails and slugs	Marine, freshwater, and terrestrial; coiled shell; prominent head with tentacles and eyes
Bivalvia — clams, oysters, and mussels	Marine, freshwater; body compressed between two hinged shells in a left and right arrangement; hatchet-shaped foot
Cephalopoda — squids and octopi	Marine; excellent swimmers, predatory; foot separated into arms and tentacles that may contain suckers; well-developed eyes

Representatives of the Phylum Brachiopoda

Classes and Representative Kinds	Characteristics
Lingulata — lamp shells	Marine; body compressed between two hinged shells in a top and bottom arrangement; stalk-like pedicle

Representatives of the Phylum Bryozoa

Representative Kinds	Characteristics
Bryozoans — moss animals	Marine, freshwater; all but one species form colonies consisting of clones called zooids; have a coelom and an internal cavity lined by mesothelium

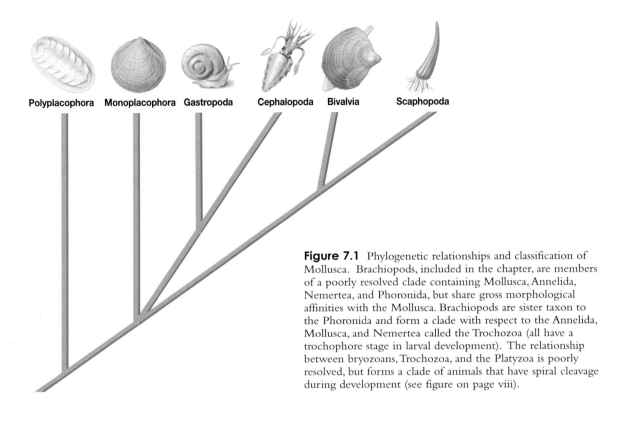

Figure 7.1 Phylogenetic relationships and classification of Mollusca. Brachiopods, included in the chapter, are members of a poorly resolved clade containing Mollusca, Annelida, Nemertea, and Phoronida, but share gross morphological affinities with the Mollusca. Brachiopods are sister taxon to the Phoronida and form a clade with respect to the Annelida, Mollusca, and Nemertea called the Trochozoa (all have a trochophore stage in larval development). The relationship between bryozoans, Trochozoa, and the Platyzoa is poorly resolved, but forms a clade of animals that have spiral cleavage during development (see figure on page viii).

Phylum Mollusca

Figure 7.2 Examples of molluscs: (a) mussels, *Mytilus californianus*, (b) cuttlefish, *Sepia bandensis*, (c) giant clam, *Tridacna derasa*, (d) sea hare, *Aplysia californica*, (e) chiton, *Nuttallina fluxa*, and (f) banana slug, *Ariolimax californicus*.

Class Polyplacophora

Figure 7.3 A spiny chiton, *Nuttallina fluxa*, at low tide.

Figure 7.4 A fuzzy chiton, *Acanthopleura granulata*, at low tide.

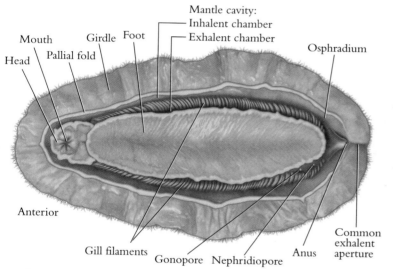

Figure 7.5 The ventral view of a chiton.

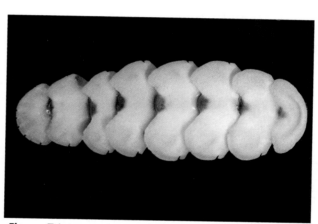

Figure 7.6 A ventral view of a chiton skeleton showing the eight dorsal plates.

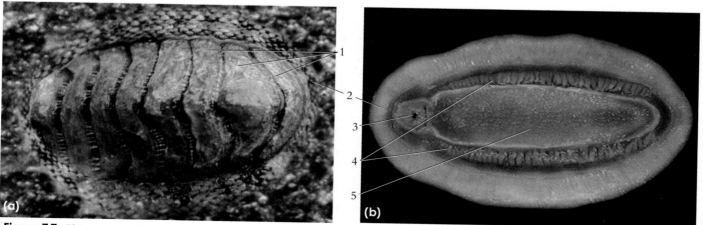

Figure 7.7 Chitons are easily recognized by their eight dorsal plates. (a) A dorsal view and (b) ventral view.
1. Dorsal plates   2. Girdle   3. Mouth   4. Gill filaments   5. Foot

Class Gastropoda

Figure 7.8 Many gastropods have ornate shells, such as the Venus comb murex, *Murex pecten* (scale in mm).

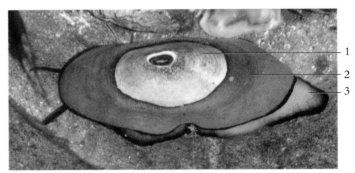

Figure 7.9 A keyhole limpet, *Megathura crenulata*.
1. Shell 2. Mantle 3. Foot

Figure 7.10 A snail, *Cornu aspersum*.
1. Shell
2. Foot
3. Occular tentacle
4. Head
5. Sensory tentacle

Figure 7.11 The locomotion of the slug, class Gastropoda, requires the production of mucus. Slugs differ from snails in that a shell is absent.
1. Foot 3. Mantle 5. Occular tentacle 7. Pneumostome
2. Mucus 4. Head 6. Sensory tentacle

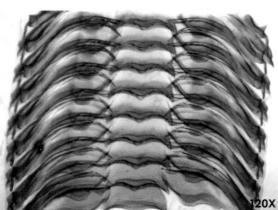

Figure 7.12 A snail radula is made up of small horny teeth made of chitin, called denticles.

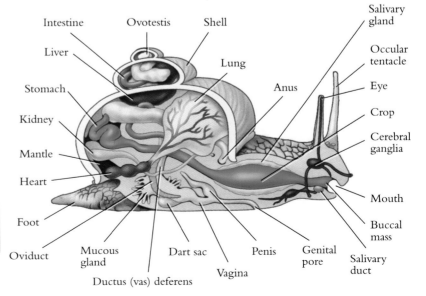

Figure 7.13 A diagram of pulmonate snail anatomy.

Class Bivalvia (=Pelycypoda)

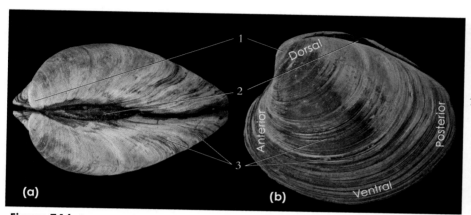

Figure 7.14 An external view of a clam shell: (a) dorsal view and (b) the left valve.
1. Umbo
2. Hinge ligament
3. Growth lines

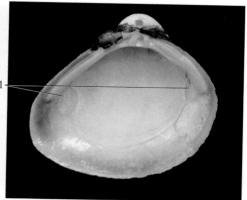

Figure 7.15 Internal view of a clam shell showing the muscle scars where the adductor muscles attached to the shell.
1. Muscle scar

Figure 7.16 A giant clam, *Tridacna derasa*.

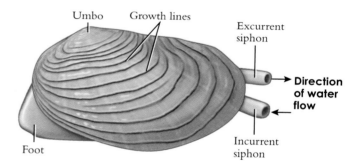

Figure 7.18 The surface anatomy of a freshwater clam, left valve.

Figure 7.17 California mussels, *Mytilus californianus*, form extensive mussel beds.

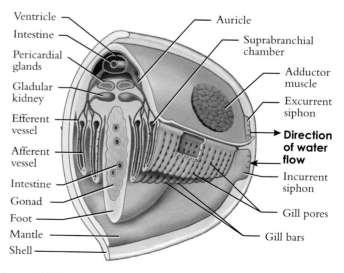

Figure 7.19 A diagram of the circulatory and respiratory systems of a freshwater clam.

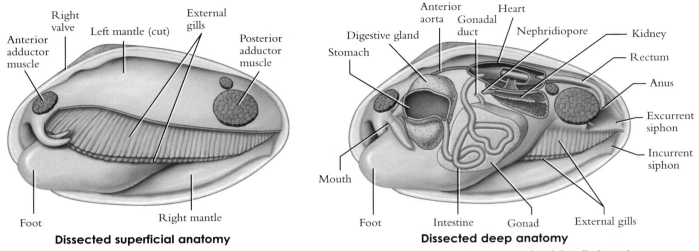

Figure 7.20 The anatomy of a freshwater clam. Bivalves have two shells (valves) that are laterally compressed and dorsally hinged.

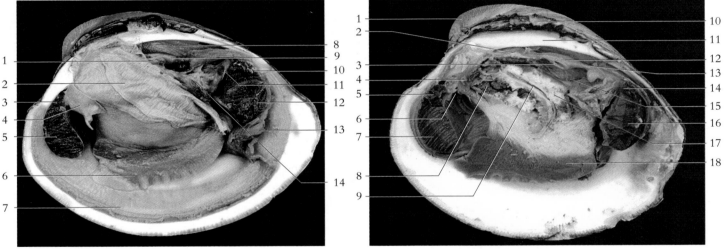

Figure 7.21 A lateral view of a clam.
1. Atrium of heart
2. Gills
3. Anterior retractor muscle
4. Labial palps
5. Anterior adductor muscle
6. Foot
7. Mantle
8. Pericardium
9. Ventricle of heart
10. Anus
11. Posterior retractor muscle
12. Posterior adductor muscle
13. Excurrent siphon
14. Nephridium (kidney)

Figure 7.22 A lateral view of a clam, foot cut.
1. Umbo
2. Intestine
3. Opening between atrium and ventricle
4. Esophagus
5. Anterior retractor muscle
6. Mouth
7. Anterior adductor muscle
8. Digestive gland
9. Intestine
10. Hinge ligament
11. Hinge
12. Ventricle of heart
13. Posterior aorta
14. Posterior retractor muscle
15. Nephridium (kidney)
16. Posterior adductor muscle
17. Gonad
18. Foot

Class Cephalopoda

Figure 7.23 Example cephalopods, (a) the giant octopus, *Enteroctopus* sp., (b) cuttlefish, *Sepiidae* sp., and (c) nautilus, *Nautilus pompilius*.

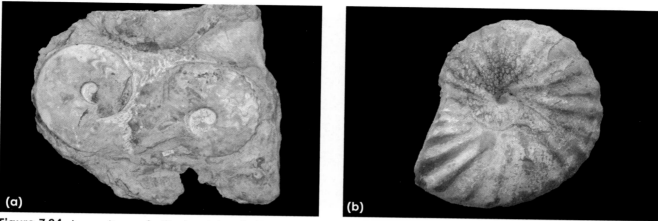

Figure 7.24 Ammonites are fossilized cephalopods. (a) Ammonite fossil encrusted limestone, and (b) *Metoicoceras geslinianum*, a late cretaceous ammonite.

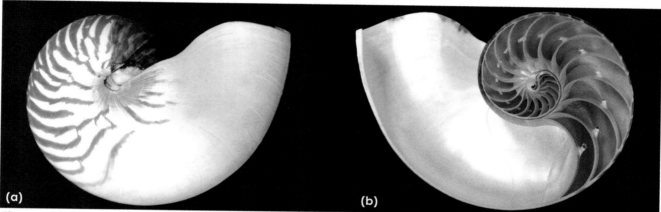

Figure 7.25 The *Nautilus* sp., a cephalopod, has gas-filled chambers within its shell, as seen in this cross-section of the shell (b). These chambers regulate buoyancy.

Figure 7.26 A dorsal view of an octopus.
1. Mantle   2. Head   3. Arms

Figure 7.27 A ventral view of an octopus.
1. Suction cups   3. Mouth
2. Arm

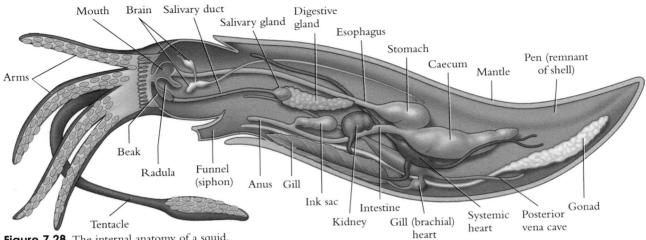

Figure 7.28 The internal anatomy of a squid.

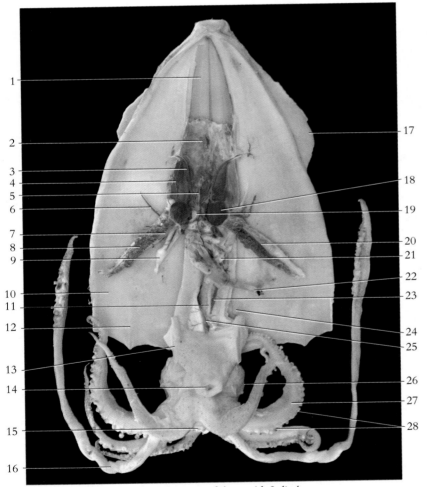

Figure 7.29 The internal anatomy of the squid, *Loligob* sp.

1. Pen (gonad partially resected)
2. Gonad
3. Lateral mantle artery
4. Posterior vena cava
5. Median mantle artery
6. Median mantle vein
7. Afferent branchial artery
8. Gill
9. Genital opening
10. Mantle
11. Esophagus
12. Articulating ridge
13. Articulating cartilage
14. Funnel (siphon)
15. Mouth
16. Tentacle
17. Fin
18. Branchial heart
19. Systemic heart
20. Efferent branchial vein
21. Ink sac
22. Rectum
23. Cephalic aorta
24. Stellate ganglion
25. Cephalic vena cava
26. Eye
27. Arm
28. Suckers

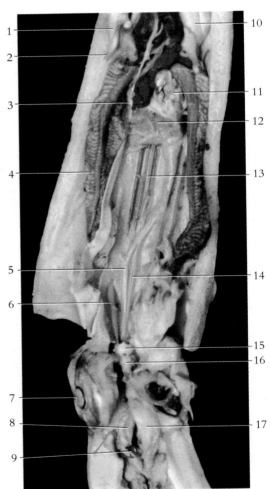

Figure 7.30 The internal anatomy of the squid, *Loligob* sp., including head region.

1. Spermatophoric duct
2. Penis
3. Kidney
4. Gill
5. Esophagus
6. Pleural nerve
7. Eye
8. Radula
9. Beak
10. Stomach
11. Pancreas
12. Digestive gland (cut)
13. Pen
14. Cephalic aorta
15. Visceral ganglion
16. Pedal ganglion
17. Buccal bulb

Mollusca, Brachiopoda, and Bryozoa

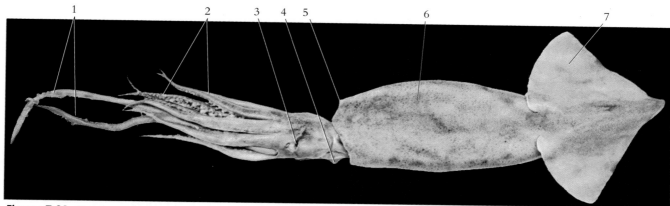

Figure 7.31 The external anatomy of the squid, *Loligob* sp.
1. Tentacles   2. Arms   3. Eye   4. Funnel (siphon)   5. Collar   6. Mantle (body tube)   7. Fin

Phylum Brachiopoda

Figure 7.32 A fossil brachiopod, *Neospirifer* sp., from the Permian period.

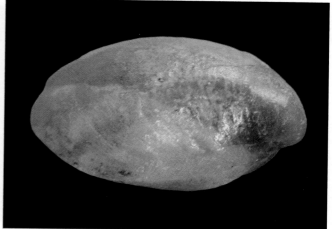

Figure 7.33 A fossil brachiopod, *Kingena* sp., from the Cretaceous period.

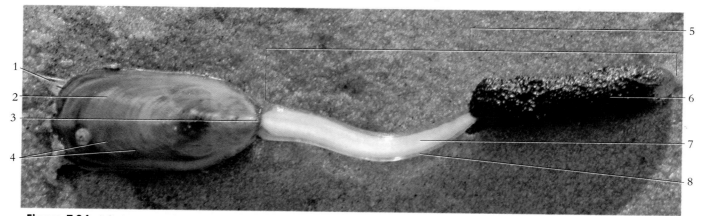

Figure 7.34 A living example of a lamp shell, *Lingula* sp.
1. Chaete   2. Pedicle valve   3. Apex   4. Growth lines   5. Pedicle   6. Substrate (sand on pedicle)   7. Muscle   8. Cuticle

Phylum Bryozoa

Figure 7.35 Illustration of the freshwater bryozoan, *Pectinatella magnifica*.

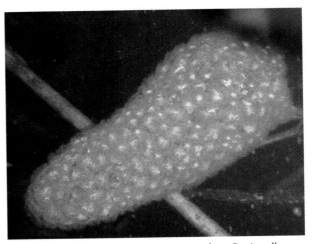

Figure 7.36 A freshwater bryozoan such as *Pectinatella magnifica* is often mistaken for a mass of frog eggs.

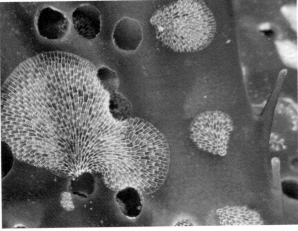

Figure 7.37 A marine bryozoan, *Membranipora membranacea*, growing on giant kelp, *Macrocystis pyrifera*.

Figure 7.38 The flute bryozoan, *Hippodiplosia insculpta*, is a marine species.

Figure 7.39 Lacy bryozoan, *Phidolopora labiata*.

Figure 7.40 A preserved bryozoan exoskeleton or zoecium.

Annelida and Nemertea

Chapter 8

An estimated 10,000 species are contained within the phylum *Annelida*. Annelids are marine, freshwater, and burrowing terrestrial, body-segmented worms that include such forms as sandworms, earthworms, and leeches. The annelid body contains a true coelom, a tubular digestive tract extending from the mouth to the anus, cerebral ganglia, a closed circulatory system with a series of hearts, and a hydrostatic skeleton with accompanying circular and longitudinal muscles for locomotion. Locomotion is aided in all annelids, except the leeches, by tiny chitinous bristles called *setae*. The stiff setae of burrowing annelids also aid in preventing them from being pulled out or washed out of the soil.

Sexes are separate or have both male and female organs in the same organism (*hermaphrodite*). Some of the annelids have the capability of regeneration, with each severed portion being capable of regenerating a complete organism.

Annelids are extremely important in the food chain and the general community ecology. Earthworms aerate and enrich the soil. Most of the leeches are predators, some are temporary parasites, and a few are permanent parasites.

Far less diverse than the annelids, nemertean species number around 1500. Most are tropical marine and burrow in sediments, but a few can be found in freshwater and even terrestrial ecosystems. Some nemerteans are scavengers, but most are carnivorous, using their evertible proboscis to capture their prey (primarily mollusks, annelids, and arthropods). In addition to their proposcis, all nemerteans also have a rhynchocoel, which is the tubular cavity that holds the inverted proboscis. It is a true coeom, lined by epithelium. The outermost layer of the body consists of a ciliated epithelium and mucus glands. Almost all species have separate sexes, although some of the larger species can be broken up into fragments each of which are capable of growing to adulthood. The vast majority of nemerteans are very small and thin, but at least one specimen measured 54 meters (177 feet) in length, making it the longest animal on Earth.

Table 8.1 Some Representatives of the Phylum Annelida

Classes and Representative Kinds	Characteristics
Polychaeta — tubeworms and sandworms	Mostly marine; segments with parapodia
Clitellata (subclass Oligochaeta) — earthworms	Freshwater and burrowing terrestrial forms; small setae; poorly developed head
Clitellata (subclass Hirudinea) — leeches	Freshwater; some are blood-sucking parasites and others are predators; lack setae; prominent muscular suckers

Some Representatives of the Phylum Nemertea

Classes and Representative Kinds	Characteristics
Anopla — proboscis worms	Mostly marine, but some freshwater and terrestrial; use evertible proboscis to catch prey and feed

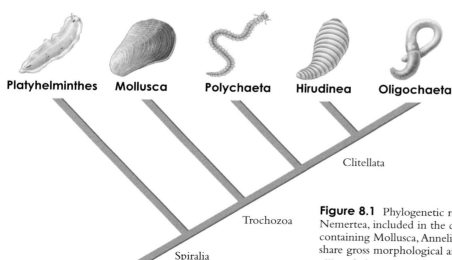

Figure 8.1 Phylogenetic relationships and classification of Annelida. Nemertea, included in the chapter, are members of a poorly resolved clade containing Mollusca, Annelida, Nemertea, Brachiopoda, and Phoronida, but share gross morphological affinities with the Annelida (see figure on page viii—phylogeny and classification of Metazoa (multicellular animals).

Phylum Annelida

Figure 8.2 Examples of annelids: (a) a tubeworm, *Protula magnifica*, (b) a bloodworm, *Glycera americana*, (c) a lugworm, *Arenicola marina*, (d) a leech, *Myzobella* sp., (e) an earthworm, *Lumbricus* sp. and (f) a sandworm, *Nereis virens*.

Class Polychaeta

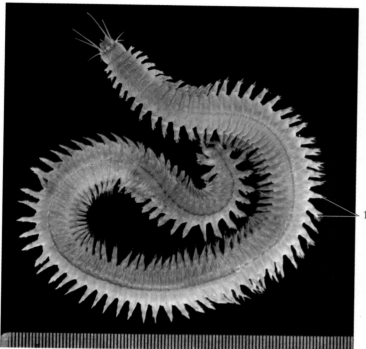

Figure 8.3 The sandworm, *Nereis virens* (scale in mm).
1. Parapodia

Figure 8.4 The anterior end of the sandworm, *Nereis virens*. (a) A dorsal view and (b) a ventral view.
1. Palpi
2. Prostomium
3. Peristomial cirri (tentacles)
4. Peristome
5. Parapodia
6. Setae
7. Mouth
8. Everted pharynx

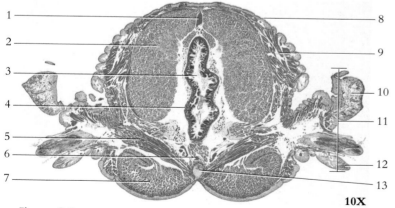

Figure 8.5 A transverse section of the sandworm, *Nereis* sp.
1. Dorsal blood vessel
2. Dorsal longitudinal muscle
3. Lumen of intestine
4. Intestine
5. Oblique muscle
6. Ventral blood vessel
7. Ventral longitudinal muscle
8. Integument
9. Circular muscle
10. Notopodium
11. Parapodium
12. Neuropodium
13. Ventral nerve cord

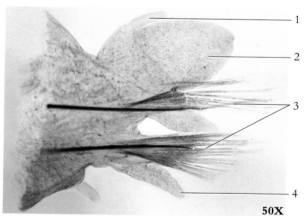

Figure 8.6 The parapodium of the sandworm, *Nereis* sp.
1. Dorsal cirrus
2. Notopodium
3. Setae
4. Neuropodium

Class Clitellata (subclass Oligochaeta)

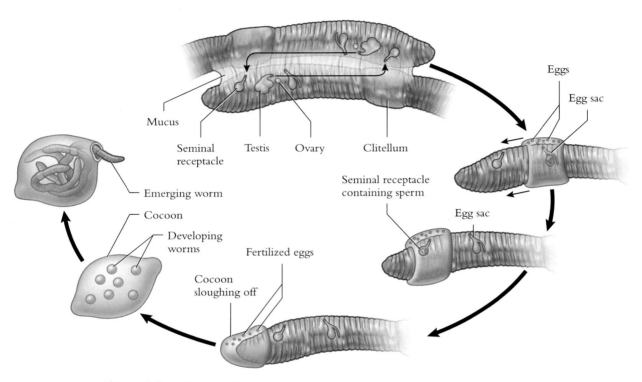

Figure 8.7 A diagram of earthworm copulation and the formation of an egg cocoon.

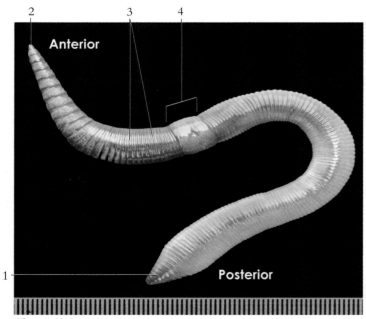

Figure 8.8 A dorsal view of an earthworm, *Lumbricus* sp. (scale in mm).

1. Pygidium
2. Prostomium (located dorsal to mouth)
3. Segments, or metameres
4. Clitellum

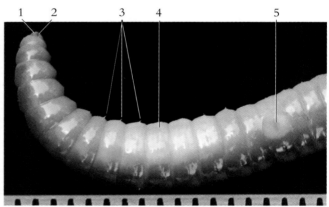

Figure 8.9 An anterior end of an earthworm, *Lumbricus* sp. (scale in mm).

1. Prostomium
2. Mouth
3. Setae
4. Segment 10
5. Opening of ductus (vas) deferens

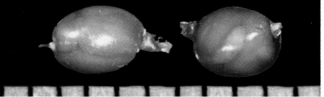

Figure 8.10 Earthworm cocoons (scale in mm).

Class Clitellata (subclass Hirudinea)

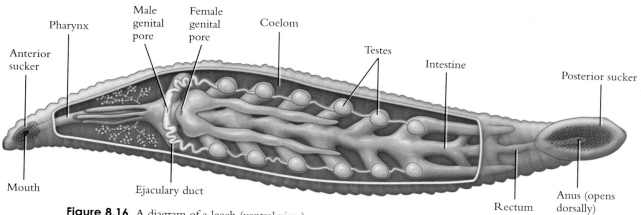

Figure 8.16 A diagram of a leech (ventral view).

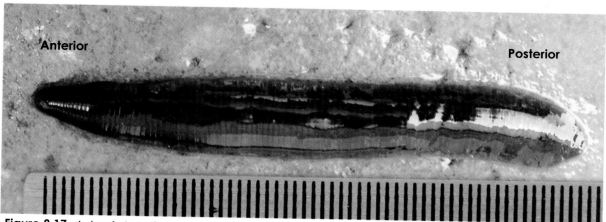

Figure 8.17 A dorsal view of a leech, *Myzobella* sp. Leeches are more specialized than other annelids. They have lost their setae and developed suckers for attachment while sucking blood (scale in mm).

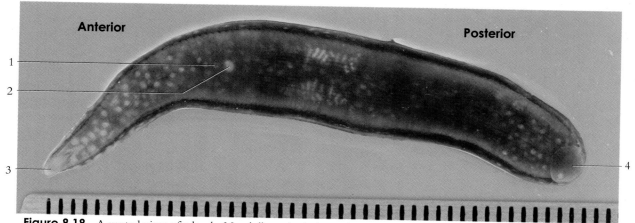

Figure 8.18 A ventral view of a leech, *Myzobella* sp. (scale in mm).
1. Male genital pore 2. Female genital pore 3. Anterior sucker 4. Posterior sucker

Phylum Nemertea

Class Anopla

Figure 8.19 *Parboriasia corrugatus*, a large nemertean from the Ross Sea, Antarctica. Typical nemertean cuticle is covered with mucus glands, especially at the anterior end, and they use their external cilia and muscular peristaltic undulation to glide on their trails of slime.

Figure 8.20 The zebra ribbon worm, *Baseodiscus mexicanus*, feeds upon mollusks, shrimp, crabs, and small fishes. It is often mistaken for a sea snake.

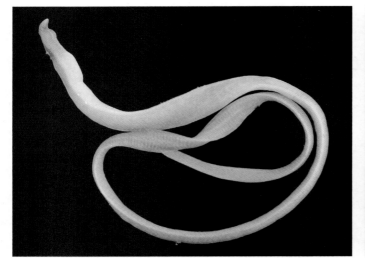

Figure 8.21 The milky ribbon worm, *Cerebratulus lacteus*.

Figure 8.22 The milky ribbon worm, *Cerebratulus lacteus*, partially buried in mud at low tide with its proboscis extended.
1. Proboscis

Nematoda and Other Pseudocoelomates

Chapter 9

An estimated 80,000 species of roundworms are contained within the phylum *Nematoda*. Nematodes include a wide variety of small, elongated animals including free-living forms in water and soil and parasitic forms such as hookworms, pinworms, *Ascaris*, and *Trichinella*.

Most of the nematodes are microscopic, bilaterally symmetrical, cylindrical, unsegmented, wormlike animals. The body of a nematode is enclosed in a tough cuticle that is shed periodically as the animal grows, an attribute shared with seven other ecdysozoan phyla. Contraction of the longitudinal skeletal muscles attached to the cuticle causes a whiplike body movement. Nematodes lack circular muscle and have an incomplete mesodermal layer. The body cavity is called a *pseudocoelom* because it lacks complete mesodermal lining. The tubular digestive tract extends the length of the body from the mouth to the anus. The sexes are usually separate, and the female is larger than the male.

It is estimated that a handful of temperate soil contains hundreds of species and many thousands of individual nematodes. They are the primary consumers of organic material and are also extremely important in subterranean and aquatic food chains. Other nematodes are parasites of plants and animals and are responsible for horrific amounts of human suffering through crop and stock animal losses. Among the human nematode parasites are the hookworms, the intestinal roundworm *Ascaris*, pinworms, trichina worms (*Trichinella*), and filarial worms.

Nematodes, along with six other phyla of animals, have a pseudocoelom and are sometimes collectively referred to as *pseudocoelomates*. Because the pseudocoelom is fluid–filled or contains a gelatinous substance, it permits a greater freedom of movement as compared to the solid body structure of acoelomates. The pseudocoelom provides a space for development and specialization of digestive, reproductive, and excretory systems.

Although the pseudoceolomates are frequently discussed together in textbooks, they are polyphyletic (not derived from a single common pseudocoelomate ancestor). The seven pseudocoelomate phyla are Nematoda, Rotifera, Gastrotricha, Kinorhyncha, Nematomorpha, Acanthocephala, and Entoprocta. Of these, the parasitic members within the phylum Nematoda are probably the most important to humans.

About 1,580 rotifer species have been described, the majority of which occupy freshwater and moist soil environments, but they are also commonly found associated with mosses and lichens. Rotifers have bilateral symmetry, a rigid outer cuticle, and a distinctive ciliated feeding structure called the corona, which they use to feed on detritus, bacteria, algae, and protozoans. They can reproduce either sexually or parthenogenetically. Members of the class Bdelloidea have undergone asexual reproduction for over 30 million years, overcoming the evolutionary problem of mutational load by taking up exogenous DNA from their own undigested food.

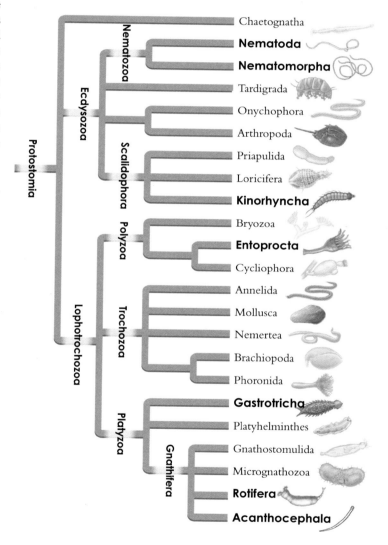

Figure 9.1 Phylogenetic relationships among protostome animals. Note the polyphyletic distribution of the pseudocoelomate phyla (Nematoda, Rotifera, Gastrotricha, Kinorhyncha, Nematomorpha, Acanthocephala and Entoprocta). Taxa in boldface are covered in this chapter).

Phylum Nematoda

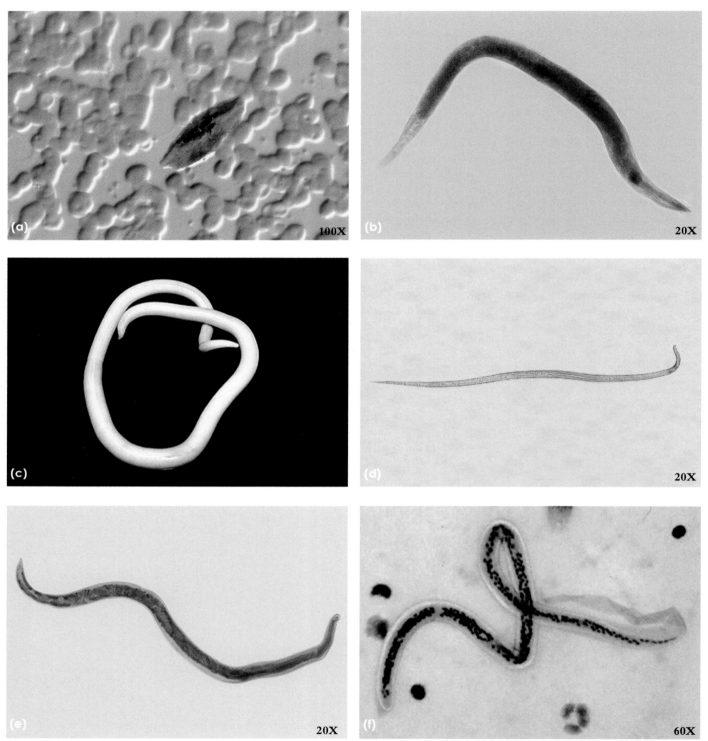

Figure 9.2 Example of Nematoda include: (a) an eye worm, *Loa loa*, in a blood smear, (b) a pinworm, *Enterobius vermicularis*, (c) an intestinal roundworm, *Ascaris* sp., (d) a vinegar eel, *Turbatrix aceti*, (e) a hookworm, *Necator americanus*, and (f) *Wuchereria bancrofti*, a parasitic nematode that if left unchecked can develop into elephantiasis.

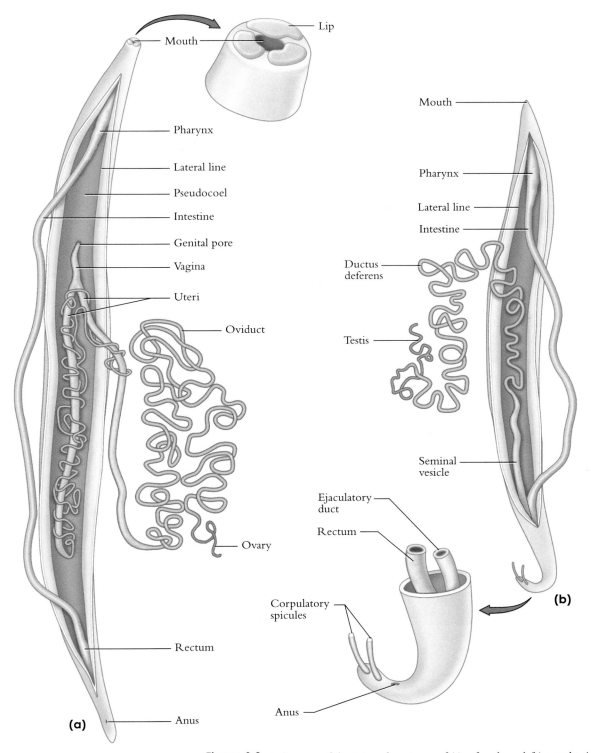

Figure 9.3 A diagram of the internal anatomy of (a) a female and (b) a male *Ascaris* sp.

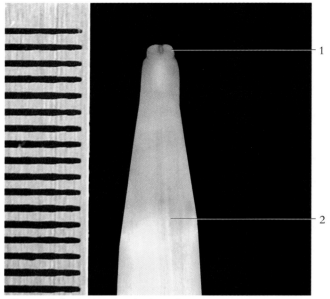

Figure 9.4 The head end of a male *Ascaris* sp. (scale in mm).
1. Lip
2. Lateral line

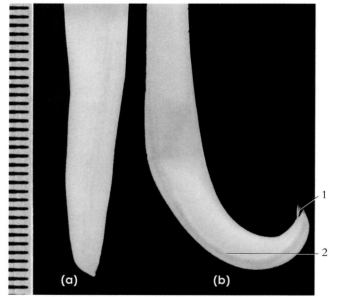

Figure 9.5 The posterior end of (a) a female and (b) a male *Ascaris* sp. (scale in mm).
1. Copulatory spicules
2. Ejaculatory duct

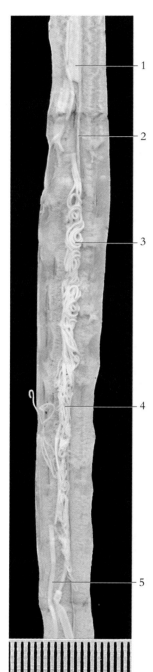

Figure 9.6 The internal anatomy of a male *Ascaris* sp. (scale in mm).
1. Intestine
2. Lateral line
3. Ductus deferens
4. Testes
5. Seminal vesicle

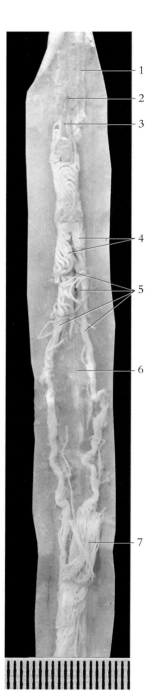

Figure 9.7 The internal anatomy of a female *Ascaris* sp. (scale in mm).
1. Intestine
2. Genital pore
3. Vagina
4. Oviducts
5. Uteri (Y-shaped)
6. Lateral line
7. Ovary

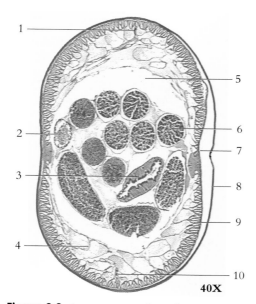

Figure 9.8 A transverse section of a male *Ascaris* sp.
1. Dorsal nerve cord
2. Ductus deferens
3. Intestine
4. Longitudinal muscle cell body
5. Pseudocoel
6. Testis
7. Lateral line
8. Cuticle
9. Contractile sheath of muscle cell
10. Ventral nerve cord

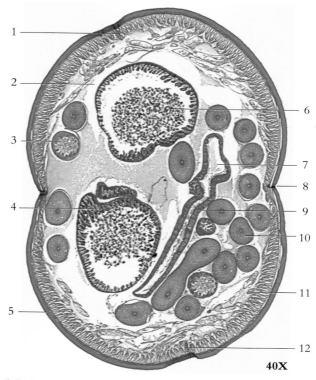

Figure 9.9 A transverse section of a female *Ascaris* sp.
1. Dorsal nerve cord
2. Pseudocoel
3. Oviduct
4. Uterus
5. Cuticle
6. Eggs
7. Lumen of intestine
8. Lateral line
9. Intestine
10. Ovary
11. Longitudinal muscles
12. Ventral nerve cord

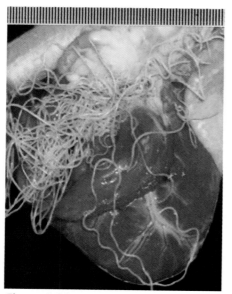

Figure 9.10 A dog heart infested with heartworm, *Dirofilaria immitis* (scale in mm).

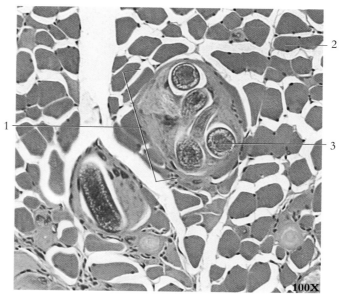

Figure 9.11 A photomicrograph of *Trichinella spiralis* encysted in muscle.
1. Cyst
2. Muscle
3. Larva of *Trichinella spiralis*

Phylum Rotifera

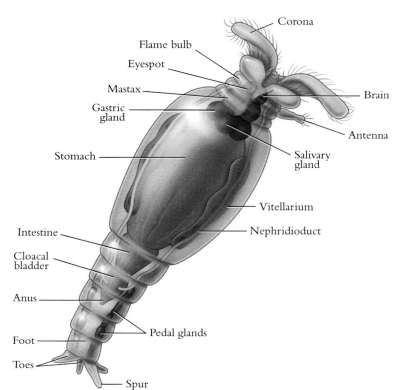

Figure 9.12 A diagram of the rotifer, *Philodina* sp.

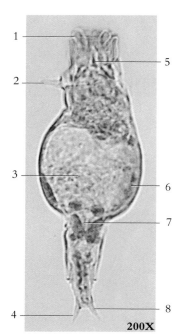

Figure 9.13 A rotifer.
1. Corona
2. Antenna
3. Stomach
4. Spur
5. Mastax
6. Vitellarium
7. Intestine
8. Toe

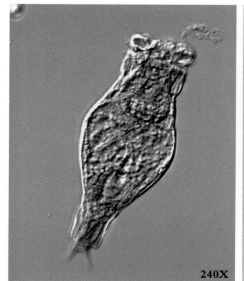 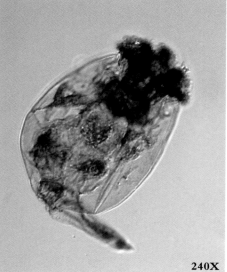 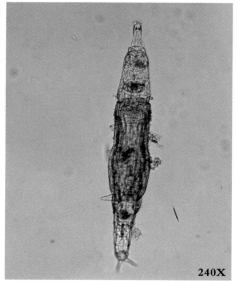

Figure 9.14 Morphological diversity of Rotifera.

Arthropoda and Tardigrada

Chapter 10

Containing over 1,235,850 species, Arthropoda is the largest phylum within the kingdom Animalia. Arthropods are also the most biologically successful of all animals. There are more species of them, they live in a greater variety of habitats, and they eat a greater variety and amount of food than the members of any other phylum. Included within this phylum are diverse organisms such as horseshoe crabs, spiders, ticks, scorpions, lobsters, crabs, shrimp, insects, centipedes, and millipedes.

Arthropods have a *segmented body*, paired and highly specialized *jointed appendages, a chitinous exoskeleton* that is periodically shed as the animal grows, and an *open circulatory system* in which the blood that is pumped by the dorsally positioned heart flows through a cavity called a *hemocoel*. The sense organs are well developed in arthropods, and most have highly specialized *compound eyes*. As compared to other invertebrates, arthropods have complex, innate (unlearned) behavior patterns.

The tremendous success of arthropods is due to the structural and physiological aspects of their body organization and their reproductive potential. Most female arthropods, especially the insects, produce thousands of eggs during their life. The eggs generally hatch when food is abundant, and the young quickly develop through gradual metamorphosis or complete metamorphosis.

The economical importance of arthropods is immeasurable. Many insects are considered pests in that they feed upon human crops. Some arthropods, such as lobsters, crabs, and shrimp, are an important source of human food. Bees pollinate flowers and produce honey for human consumption. Most spiders are considered beneficial because they feed on noxious insects. Ticks and many insects, especially flies and mosquitoes, are of medical concern because they are vectors (carriers) of pathogenic microorganisms.

Approximately 1,150 species of tardigrades have been described, and the phylum has a global distribution. Tardigrades are small, usually between 0.5 and 1.0 millimeters in length. Commonly associated with mosses and lichens, tardigrades also inhabit freshwater and marine sediments. They have four unjointed pairs of legs, but like nematodes, their mouth consists of a tri-radiate, muscular pharynx and a stylet that can be used to scrape up algal or bacterial cells or pierce the bodies of other small invertebrates, upon which they feed. Tardigrades are known to survive tremendous environmental insults, including exposure to the vacuum, temperature, desiccation, and radiation extremes of outer space.

Table 10.1 Representatives of the Phylum Arthropoda

Classes and Representative Kinds	Characteristics
Merostomata (Subphylum Chelicerata) — horseshoe crab	Cephalothorax and abdomen; specialized front appendages into chelicerae; lack antennae and mandibles
Arachnida (Subphylum Chelicerata) — spiders, mites, ticks, and scorpions	Cephalothorax and abdomen; four pairs of legs; book lungs or trachea; lack antennae and mandibles
Malacostraca (Subphylum Crustacea) — lobsters, crabs, shrimp, and isopods	Cephalothorax and abdomen; two pairs of antennae; pair of mandibles and two pairs of maxillae; biramous appendages; gills
Maxillopoda (Subphylum Crustacea) — copepods and barnacles	Cephalothorax and abdomen; freshwater and marine; up to six pairs of appendages
Insecta — beetles, butterflies, and ants	Head, thorax, and abdomen; three pairs of legs; well-developed mouth parts; usually two pairs of wings; trachea
Chilopoda — centipedes	Head with segmented trunk; one pair of legs per segment; trachea; one pair of antennae
Diplopoda — millipedes	Head with segmented trunk; usually two pairs of legs per segment; trachea

Representatives of the Phylum Tardigrada

Phylum	Characteristics
Tardigrada — water bears	Bilaterally symmetrical; four pairs of lobopod legs terminating in claws or sucking disks

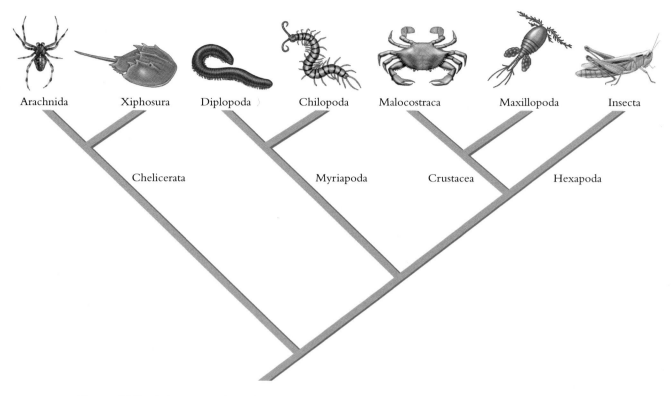

Figure 10.1 Phylogenetic relationships and classification of select arthropods (does not include classes Brachiopoda, Remipedia, Cephalocarida, and Ostracoda. Maxillopoda are likely paraphyletic). Tardigrada probably branch basally to Arthropoda and Onychophora (see figure on page viii—phylogeny and classification of Metazoa (multicellular animals).

Phylum Arthropoda

Figure 10.2 Example arthropods include: (a) flat rock scorpion, *Hadogenes troglodytes*, (b) American giant millipede, *Narceus americanus*, (c) peppermint shrimp, *Lysmata wurdemanni*, (d) tiger beetle, *Cicindela fulgida*, and (e) fossil trilobite, *Modicia typicalis*. Trilobites are extinct arthropods from the Cambrian and Ordovician periods.

Subphylum Chelicerata - Class Merostomata

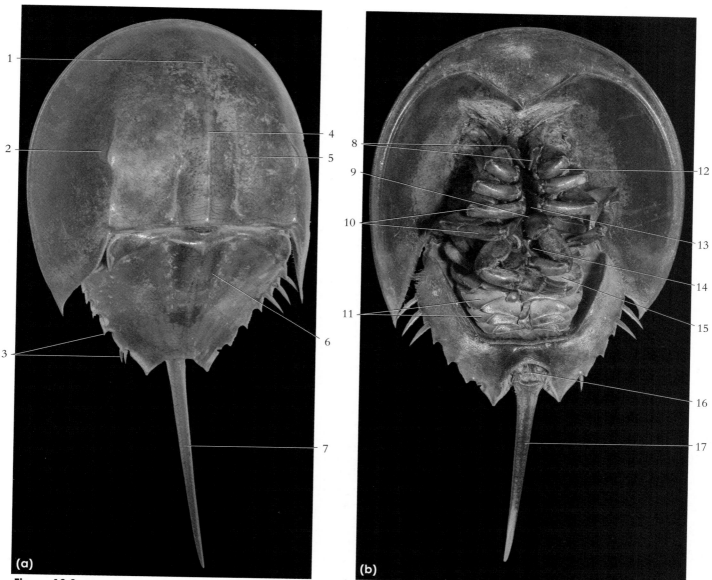

Figure 10.3 (a) A dorsal view and (b) a ventral view of the horseshoe crab, *Limulus*. This animal is commonly found in shallow waters along the Atlantic coast from Canada to Mexico.

1. Simple eye
2. Compound eye
3. Abdominal spines
4. Anterior spine
5. Cephalothorax (prosoma)
6. Abdomen (opisthosoma)
7. Telson
8. Chelicerae
9. Gnathobase
10. Chelate legs
11. Book gills
12. Pedipalp
13. Mouth
14. Chilarium
15. Genital operculum
16. Anus
17. Telson

Subphylum Chelicerata - Class Arachnida

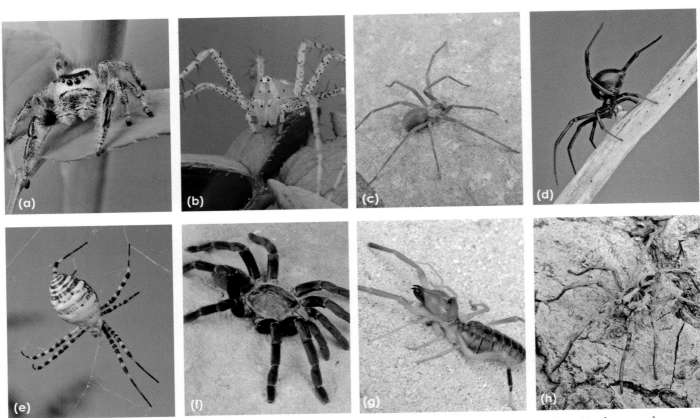

Figure 10.4 Example arachnids include: (a) a jumping spider, *Phidippus regius,* (b) a green lynx spider, *Peucetia viridans,* (c) a brown recluse, *Loxosceles apache,* (d) a black widow, *Latrodectus hesperus,* (e) an orb weaver, *Argiope trifasciata,* (f) a cobalt blue tarantula, *Haplopelma lividum,* (g) a solpugid, *Eremobates pallipes,* and (h) a wolf spider, *Lycosa* sp.

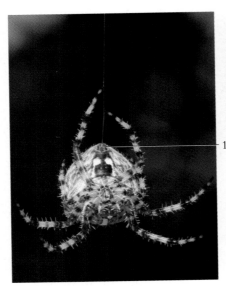

Figure 10.5 A garden spider in the process of spinning a web.
1. Spinnerets

Figure 10.6 A tick, within the family Ixodidae, is a specialized parasitic arthropod (scale in mm).

Figure 10.7 A red mite, *Dermanyssus gallinae,* feeding on a lizard.

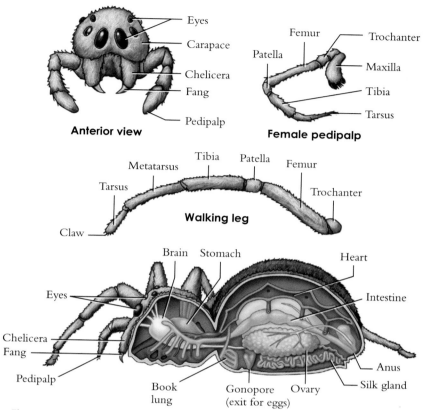

Figure 10.8 A diagram of the anatomy of a spider.

Figure 10.9 A cobalt blue tarantula, *Haplopelma lividum*.
1. Opisthosoma (abdomen)
2. Prosoma (cephalothorax)
3. Pedipalps

Figure 10.10 Examples of scorpions: (a) bark, *Centruroides hentzi*, (b) tri-colored, *Opistophthalmus ecristatus*, and (c) emperor, *Pandinus imperator*.

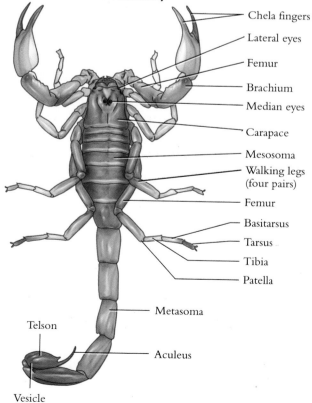

Figure 10.11 External anatomy of a scorpion.

Figure 10.12 An Arizona hairy scorpion, *Hadrurus arizonensis*. Scorpions are most commonly found in tropical and subtropical regions, but there are also several species found in arid and temperate zones.

1. Cephalothorax
2. Pedipalp
3. Stinging apparatus
4. Postabdomen (tail)
5. Preabdomen
6. Walking legs

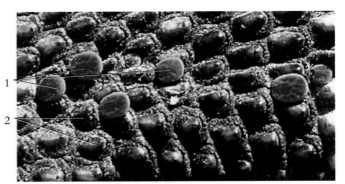

Figure 10.13 Some ticks attached and feeding on a savannah monitor, a large African lizard.

1. Ticks
2. Scales of monitor

Class Malacostraca

Figure 10.14 A pill bug, *Armadillidium* sp.

Figure 10.15 A sea slater, *Ligia italica*.

Figure 10.16 A fire shrimp, *Lysmata debelius*.

Figure 10.17 A Sally Lightfoot crab, *Grapsus grapsus*.

Figure 10.18 A hermit crab, *Coenobita clypeatus*.

Figure 10.19 The water flea, *Daphnia*, is a common microscopic crustacean.

1. Heart
2. Midgut
3. Compound eye
4. 2nd antenna
5. Rostrum
6. Setae
7. Brood chamber
8. Eggs
9. Apical spine
10. Hindgut
11. Abdominal setae
12. Anus
13. Abdominal claw
14. Carapace

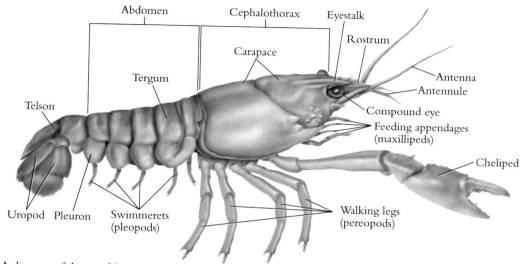

Figure 10.20 A diagram of the crayfish, *Cambarus*.

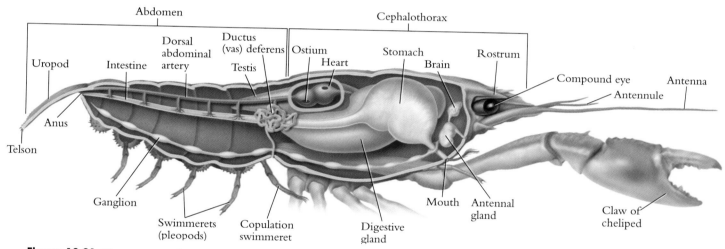

Figure 10.21 The anatomy of a crayfish. A sagittal section of an adult male.

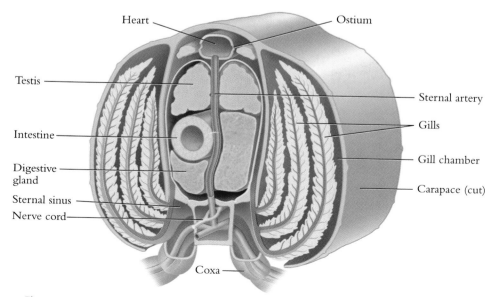

Figure 10.22 The anatomy of a crayfish. A transverse section of an adult male.

A Photographic Atlas for the Zoology Laboratory

Figure 10.23 A dorsal view of the crayfish.
1. Cheliped
2. Walking legs
3. Carapace
4. Abdomen
5. Telson
6. Uropod
7. Antenna
8. Antennule
9. Rostrum
10. Compound eye
11. Cephalothorax
12. Tergum

Figure 10.24 A lateral view of the crayfish.
1. Carapace
2. Abdomen
3. Uropod
4. Swimmeret (pleopod)
5. Rostrum
6. Compound eye
7. Maxilliped
8. Cheliped
9. Walking legs

Figure 10.25 A ventral view of the oral region of the crayfish.
1. Third maxilliped
2. Mandible
3. Second maxilla
4. Green gland duct
5. First maxilliped

Arthropoda and Tardigrada

Figure 10.26 A dorsal view of the oral region of the crayfish.
1. Compound eye
2. Walking leg
3. Green gland
4. Cardiac chamber of stomach
5. Brain
6. Circumesophageal connection (of ventral nerve cord)
7. Esophagus
8. Region of gastric mill
9. Digestive gland
10. Gill

Figure 10.27 A dorsal view of the anatomy of a crayfish.
1. Antenna
2. Compound eye
3. Brain
4. Circumesophageal connection (of ventral nerve cord)
5. Mandibular muscle
6. Digestive gland
7. Gills
8. Antennules
9. Walking legs
10. Green gland
11. Esophagus
12. Pyloric stomach
13. Testis
14. Ductus deferens
15. Aorta
16. Intestine

Figure 10.28 A ventral view of (a) a female and (b) a male crayfish. The first pair of swimmerets are greatly enlarged in the male for the depositing of sperm in the female's seminal receptacle.
1. Third maxilliped
2. Walking legs
3. Disk covering oviduct
4. Seminal receptacle
5. Abdomen
6. Base of cheliped
7. Base of last walking leg
8. Copulatory swimmerets (pleopods)
9. Sperm ducts (genital pores)
10. Swimmerets (pleopods)

Class Insecta

Figure 10.29 Example insects include: (a) a greater arid-land katydid, *Neobarrettia spinosa*, (b) an Eastern lubber grasshopper, *Romalea microptera*, (c) a flame skimmer dragonfly, *Libellula saturata*, (d) a cicada, *Diceroprocta apache*, (e) a milkweed beetle, *Tetraopes tetraophthalmus*, (f) a cynthia moth, *Samia cynthia*, (g) a giant cockroach, *Blaberus giganteus*, and (h) a Carolina mantis, *Stagmomantis carolina*.

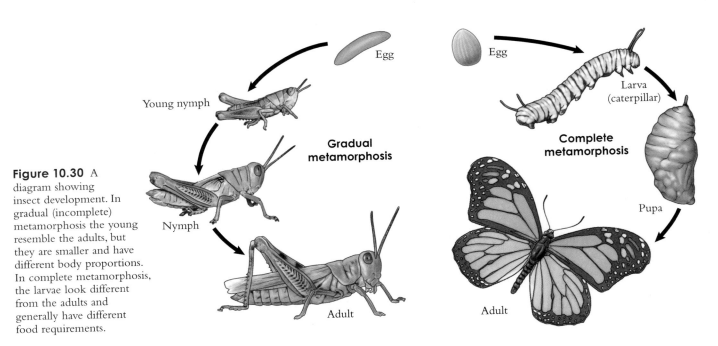

Figure 10.30 A diagram showing insect development. In gradual (incomplete) metamorphosis the young resemble the adults, but they are smaller and have different body proportions. In complete metamorphosis, the larvae look different from the adults and generally have different food requirements.

  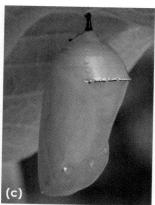 

Figure 10.31 The developmental stages of the monarch butterfly, *Danaus plexippus*, include (a) egg, (b) larval stage, (c) chrysalis, and (d) adult.

Figure 10.32 The pupa of the Ailanthus silkmoth, *Samia cynthia*. The silken cocoon has been removed (scale in mm).

Figure 10.33 A grasshopper, nymph, *Melanoplus*.

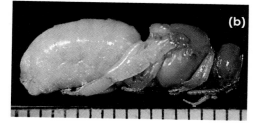

Figure 10.34 A common house cricket, *Acheta domestica*, molting. All arthropods must periodically shed their exoskeleton in order to grow. This process is called molting, or ecdysis.

Figure 10.35 The developmental stages of the common honeybee, *Apis mellifera*, include (a) larval stage, (b) pupa, and (c) adult (scale in mm).

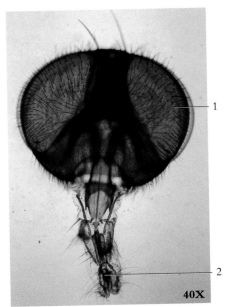

Figure 10.36 The head of a housefly, showing an example of a sponging type mouthpart in insects. Notice the large lobes at the apex of the labium, which function in lapping up liquids.
1. Compound eye
2. Labium

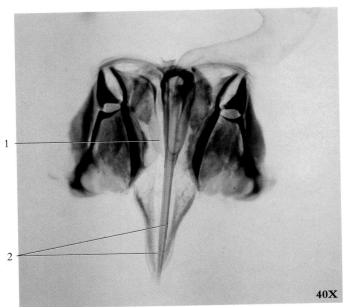

Figure 10.37 A honeybee stinger. The two darts contain barbs on the tips that point upward, making it difficult to remove a stinger from a wound.
1. Sheath
2. Darts

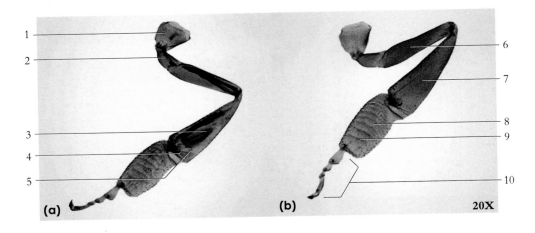

Figure 10.38 The hind legs of a worker honeybee, *Apis mellifera*, (a) outer surface and (b) inner surface.
1. Coxa
2. Trochanter
3. Pollen basket
4. Pollen packer
5. Pecten
6. Femur
7. Tibia
8. Metatarsus
9. Pollen comb
10. Tarsus

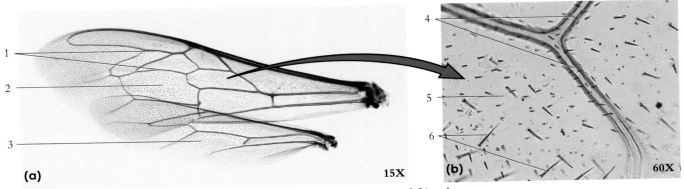

Figure 10.39 The wings of the honeybee, *Apis mellifera*. (a) A whole mount and (b) a close-up.
1. Cross veins
2. Forewing
3. Hindwing
4. Cross veins
5. Transparent wing film
6. Hairs

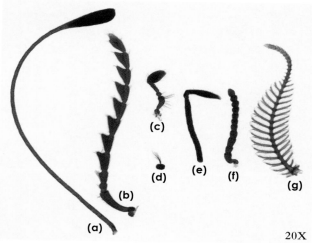

Figure 10.40 Some common insect antennae. (a) Clavate—butterflies, (b) serrate—click beetles, (c) lamellate—scarab beetles, (d) aristate—houseflies, (e) geniculate—weevils (f) moniliform—termites, and (g) plumose—moths.

Figure 10.41 The plumose antennae of the Ailanthus silkmoth, *Samia cynthia*.

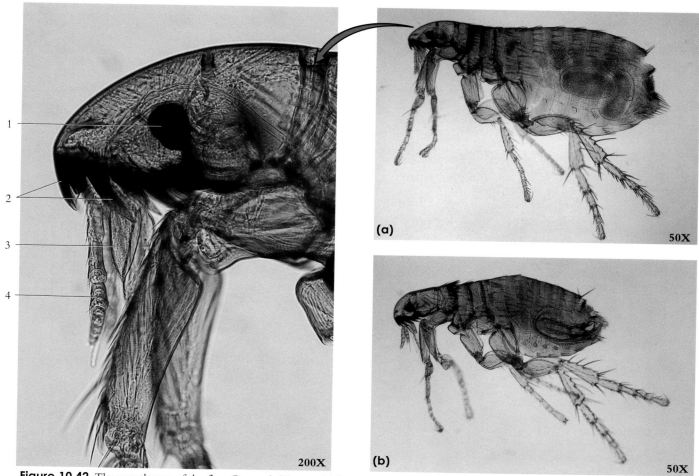

Figure 10.42 The mouthparts of the flea, *Ctenocephalide* sp., which are specialized for parasitism. Notice the oral bristles beneath the mouth that aid the flea in penetrating between hairs to feed on the blood of mammals. (a) Female flea and (b) male flea.

1. Eye
2. Oral bristles
3. Maxilla
4. Maxillary palp

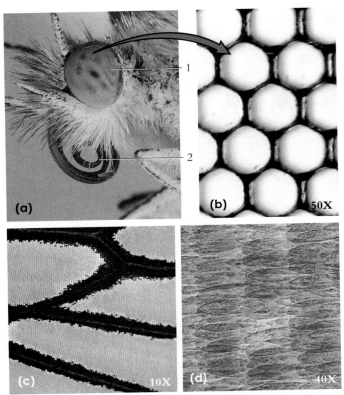

Figure 10.43 (a) A lateral view of the head of a butterfly. The most obvious structures on the head of a butterfly are compound eyes and the curled tongue for siphoning nectar from flowers. (b) A magnified view of the compound eye. (c) A close-up view of the wing scales and (d) a magnified view of the wing scales.

1. Compound eye
2. Tongue

Figure 10.44 Anatomy of the grasshopper. (a) Male and (b) female.

1. Antenna
2. Ocelli
3. Compound eye
4. Prothorax
5. Mesothorax
6. Tympanum
7. Femur
8. Pronotum
9. Mandible
10. Labrum
11. Labial palp
12. Metathorax
13. Tibia
14. Cercus
15. Subgenital plate
16. Spiracle
17. Tarsus
18. Tegmen
19. Wing
20. Abdomen
21. Dorsal valve
22. Ventral valve
23. Ovipositor

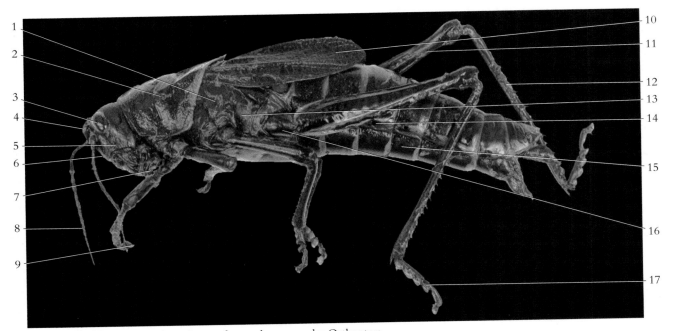

Figure 10.45 A preserved specimen of a grasshopper, order Orthoptera.

1. Mesothorax
2. Pronotum
3. Compound eye
4. Vertex
5. Gena
6. Frons
7. Maxilla
8. Antenna
9. Claw
10. Wing
11. Femur
12. Tibia
13. Mesothorax
14. Spiracle
15. Abdomen
16. Trochanter
17. Tarsus

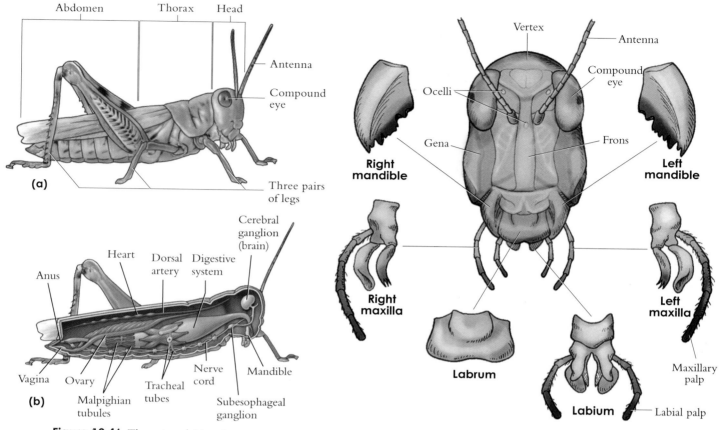

Figure 10.46 The external (a) and internal (b) anatomy of a grasshopper.

Figure 10.47 A diagram of the head and mouthparts of a grasshopper.

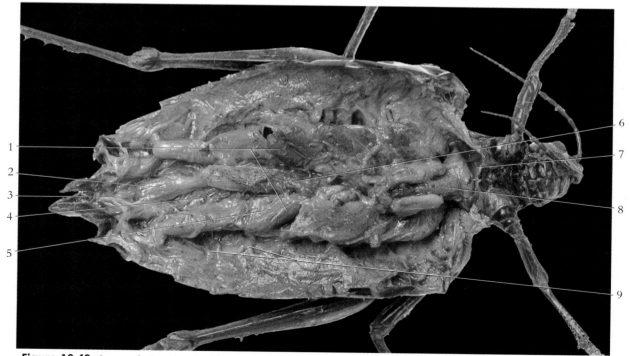

Figure 10.48 A ventral view showing the internal anatomy of a grasshopper.

1. Gastric caecum
2. Hindgut
3. Rectum
4. Malpighian tubules
5. Ovaries
6. Midgut
7. Esophagus
8. Crop
9. Tracheae

Class Chilopoda

Figure 10.49 Example centipedes include: (a) a giant Sonoran, *Scolopendra heros,* (b) a Florida blue, *Hemiscolopendra marginata*, and (c) a Vietnamese centipede, *Scolopendra subspinipes*.

Class Diplopoda

  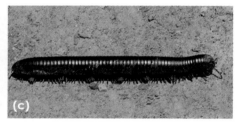

Figure 10.50 Example millipedes include: (a) an American giant millipede, *Narceus americanus*, (b) a Sonoran desert, *Orthoporus ornatus*, and (c) an African giant millipede, *Archispirostreptus gigas*.

Phylum Tardigrada

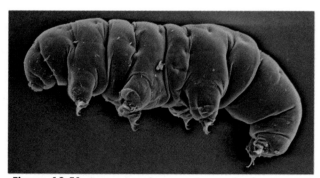

Figure 10.51 A scanning electron micrograph of a eutardigrade. Lateral view, anterior end is to the left.

Figure 10.52 A scanning electron micrograph of a eutardigrade. Ventral view, anterior is to the right.

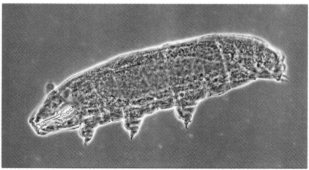

Figure 10.53 A light micrograph of *Macrobiotus polaris*. A lateral view with the anterior end to the left.

Figure 10.54 A scanning electron micrograph of a heterotardigrade. Ventral view, anterior is to the top.

Echinodermata and Hemichordata

Chapter 11

An estimated 7,510 species of extant echinoderms are contained within the phylum Echinodermata. Echinoderms are bottom-dwelling marine animals that include such forms as sea stars (starfish), sea urchins, sand dollars, sea cucumbers, and crinoids.

Larval echinoderms have bilateral symmetry, but the adults possess five-sided, or pentaradial symmetry. Echinoderms have a well-developed coelom; most have a complete digestive system, a rudimentary circulatory system, a simple nervous system consisting of nerve rings surrounding the mouth with radiating nerves, and various types of respiratory structures (e.g., dermal gills in sea stars; respiratory trees in sea cucumbers). In some, the endoskeleton is of small calcareous plates that bear outwardly projecting spines. Echinoderms have a water-vascular system composed of a network of canals through which seawater circulates. Branches of this system lead to numerous tube feet that extend when filled with water. The tube feet function in locomotion, obtaining food, and in some forms, gas exchange. Most echinoderms have regenerative capacities, and an entire sea star can regenerate from a single severed arm.

Echinoderms are extremely important in the marine food chain. Sea urchins are food for many organisms, ranging from sea stars to sea otters. Many people in Asia consume large quantities of sea cucumbers and the eggs of sea urchins. Because sea stars commonly feed on clams, oysters, and other mollusks, they are considered a threat to the shellfish industry.

Echinoderms are also important laboratory animals used in developmental biology experiments. Their gametes are easily harvested and are sufficiently large that they can be readily studied.

There are about 120 extant species of Hemichordata, and the current best estimate of phylogenetic relationships suggests that they are most closely related to echinoderms. Hemichordate larvae are very similar to echinoderm larvae, but later stages of development share several features with the chordates, including pharyngeal gill slits and a dorsal hollow nerve cord. Hemichordates possess a stomochord that looks like a chordate notochord, but that was probably derived independently through convergent evolution. Their shared features with chordates make them an important taxon for understanding the origin and evolution of the phylum Chordata.

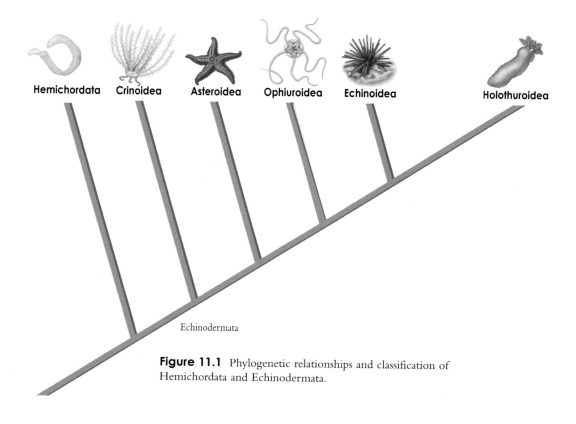

Figure 11.1 Phylogenetic relationships and classification of Hemichordata and Echinodermata.

Table 11.1 Representatives of the Phylum Echinodermata

Classes and Representative Kinds	Characteristics
Crinoidea — sea lilies and feather stars	Sessile during much of life cycle; calyx supported by elongated stalk in some
Asteroidea — sea stars (starfish)	Appendages arranged around a central disk containing the mouth; tube feet with suckers
Echinoidea — sea urchins, and sand dollars	Disk-shaped with no arms; skeleton consists of rows of calcium carbonate plates; movable spines; tube feet with suckers
Ophiuroidea — brittle stars	Appendages sharply marked off from central disk; tube feet without suckers
Holothuroidea — sea cucumbers	Cucumber-shaped with no arms; spines absent; tube feet with tentacles and suckers

Representatives of the Phylum Hemichordata

Classes and Representative	Characteristics
Enteropneusta — acorn worm	Vermiform with acorn-shaped proboscis; skin covered with cilia and mucus glands; feed on detritus by swallowing sediment
Pterobranchia — gill-wing worms	Colonial; tentacles with cilia filter food from water

Phylum Echinodermata

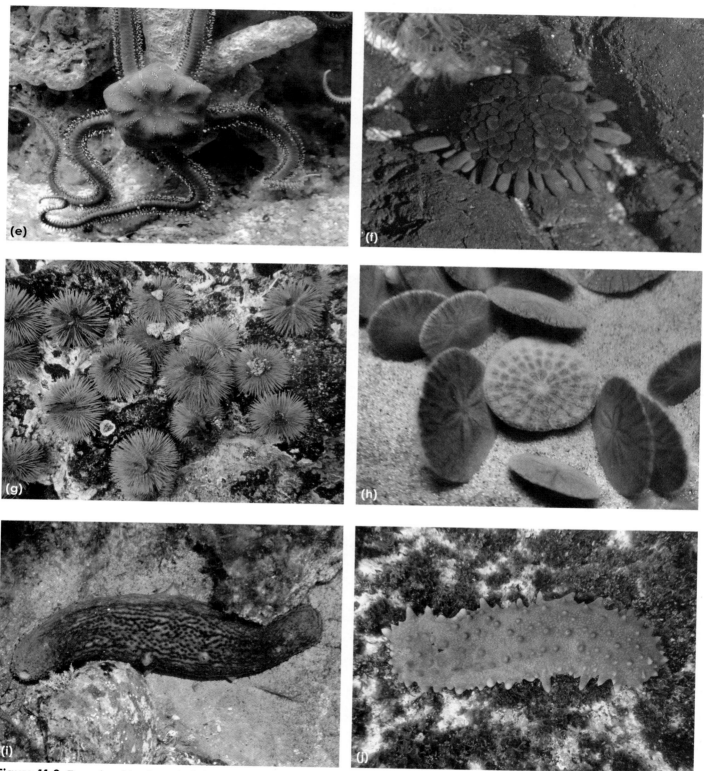

Figure 11.2 Example echinoderms include (starting on previous page): (a) a red stalked crinoid, subclass *Articulata*, (b) a group of sea stars (starfish), *Asterias* sp., in a tide pool, (c) a chocolate chip sea star (starfish), *Protoreaster* sp., (d) a yellow pyramid sea star (starfish), *Pharia Pyramidata*, (e) a green brittle star, *Ophiarachna incrassata*, (f) a helmet sea urchin, *Colobocentrotus atratus*, (g) a group of green sea urchins, *Strongylocentrotus droebachiensis*, (h) common sand dollars, *Echinarachnius parma*, (i) a California sea cucumber, *Parastichopus californicus*, and (j) a Galapagos sea cucumber, *Stichopus fuscus*.

Class Asteroidea

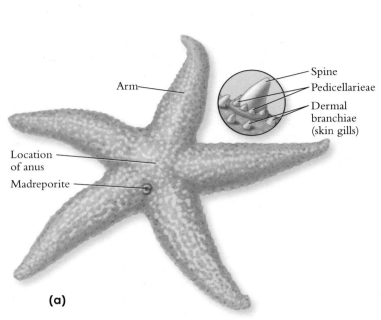

Figure 11.3 A diagram of the external anatomy of the sea star (starfish), *Asterias*. (a) An aboral (dorsal) view and (b) an oral (ventral) view.

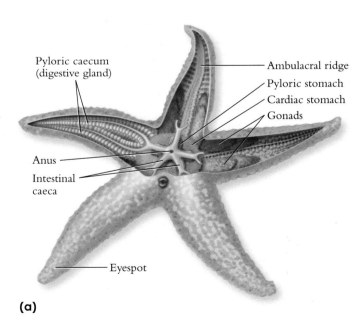

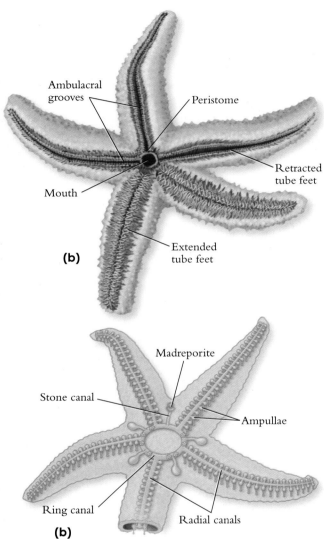

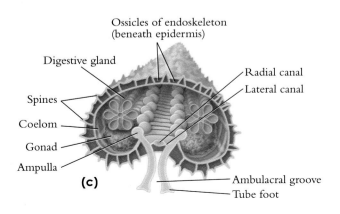

Figure 11.4 A diagram of the internal anatomy of the sea star from aboral side. (a) The digestive and reproductive organs, (b) the water vascular system, and (c) a transverse section through an arm.

Echinodermata and Hemichordata

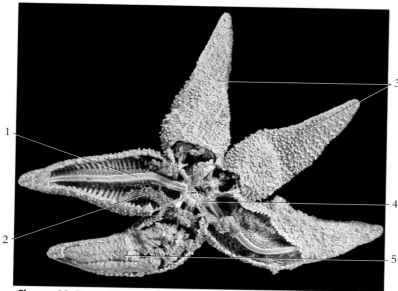

Figure 11.5 An aboral view of the internal anatomy of a sea star.
1. Ambulacral ridge
2. Gonad
3. Spines
4. Ring canal
5. Pyloric caecum (digestive gland)

Figure 11.6 A magnified aboral view of the internal anatomy of a sea star.
1. Ambulacral ridge
2. Madreporite
3. Stone canal
4. Ampullae
5. Polian vesicle
6. Pyloric duct
7. Pyloric caecum (digestive gland)
8. Gonad
9. Anus
10. Pyloric stomach
11. Spines

Figure 11.7 An oral view of a sea star.
1. Tube feet
2. Peristome
3. Mouth
4. Ambulacral groove
5. Oral spines

Figure 11.8 A transverse section through the arm of a sea star.
1. Coelom
2. Tube foot
3. Epidermis
4. Sucker
5. Pyloric caecum
6. Ambulacral ridge
7. Ampullae
8. Ambulacral groove

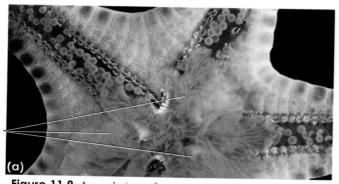
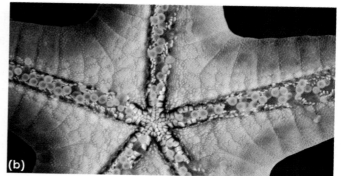

Figure 11.9 An oral view of a sea star (a) showing the cardiac stomach extended through mouth and (b) after retracting stomach.
1. Cardiac stomach

Class Echinoidea

Figure 11.10 A pencil sea urchin, *Heterocentrotus* sp.

Figure 11.11 An oral view of a live sea urchin.
1. Tips of teeth (of Aristotle's lantern)
2. Mouth
3. Peristome

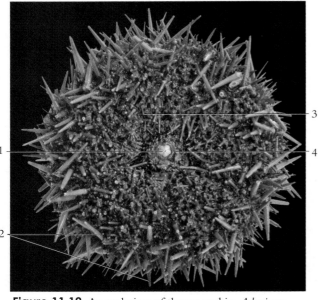

Figure 11.12 An oral view of the sea urchin, *Arbacia* sp.
1. Tip of teeth (of Aristotle's lantern)
2. Spines
3. Pedicellaria
4. Peristome

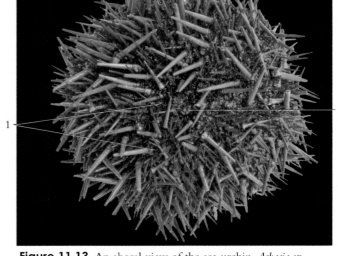

Figure 11.13 An aboral view of the sea urchin, *Arbacia* sp.
1. Ossicles
2. Madreporite

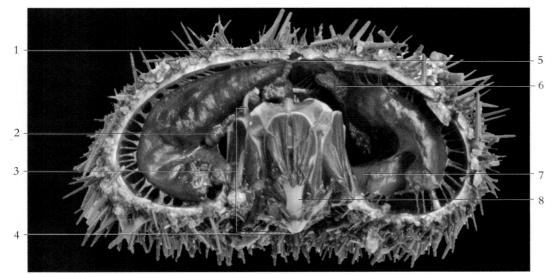

Figure 11.14 The internal anatomy of a sea urchin.
1. Madreporite
2. Intestine
3. Aristotle's lantern
4. Tip of teeth (of Aristotle's lantern)
5. Anus
6. Gonad
7. Stomach
8. Calcareous tooth

Class Holothuroidea

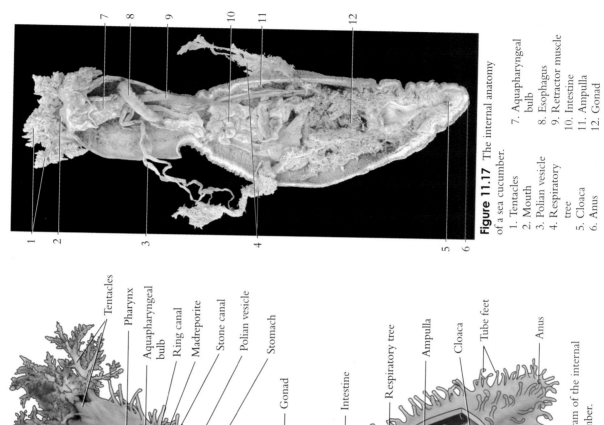

Figure 11.17 The internal anatomy of a sea cucumber.
1. Tentacles
2. Mouth
3. Polian vesicle
4. Respiratory tree
5. Cloaca
6. Anus
7. Aquapharyngeal bulb
8. Esophagus
9. Retractor muscle
10. Intestine
11. Ampulla
12. Gonad

Figure 11.16 A diagram of the internal anatomy of a sea cucumber.

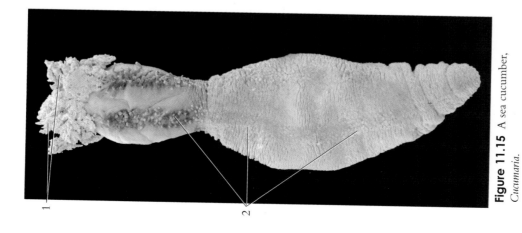

Figure 11.15 A sea cucumber, *Cucumaria*.
1. Tentacles
2. Tube feet

Phylum Hemichordata

Class Enteropneusta

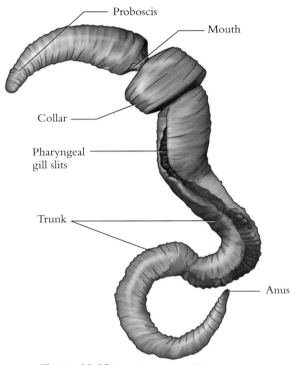

Figure 11.18 An illustration of an acorn worm, *Saccoglossus kowalevskii*.

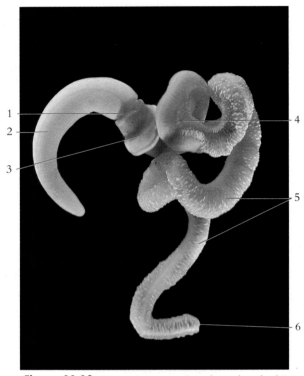

Figure 11.19 An acorn worm, *Saccoglossus kowalevskii*.
1. Mouth
2. Proboscis
3. Collar
4. Location of gills
5. Trunk
6. Anus

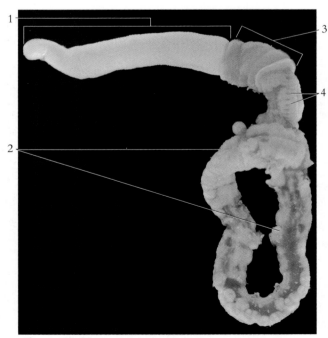

Figure 11.20 An acorn worm, *Saccoglossus kowalevskii*, showing pharyngeal gill slits.
1. Proboscis
2. Trunk
3. Collar
4. Pharyngeal gill slits

Figure 11.21 A deep-sea acorn worm species newly discovered by NOAA (National Oceanic and Atmospheric Administration).

Chordata

Chapter 12

An estimated 67,600 species of chordates are contained within the phylum *Chordata*. Chordates include a wide variety of animals that have, during some stage of development, a notochord, pharyngeal gill slits, dorsal hollow nerve cord, a muscular post-anal tail, and an endostyle, which is a groove in the ventral wall of the pharynx. Included in this phylum are the tunicates, the lancelets, and the vertebrate animals.

Chordates are thought to have descended from echinoderm-like ancestors during the Precambrian period. Tunicates are small, primitive chordates that are filter-feeders of plankton within marine waters. Lancelets are small fish-like chordates that inhabit sandy bottoms of coastal waters. The five chordate characteristics are present and functional in adult lancelets. Vertebrates are chordates that have a well-developed brain and a cartilaginous or bony vertebral column surrounding a dorsal nerve cord. There are an estimated 43,000 species of living vertebrates within eight classes —*Agnatha* (jawless fishes), *Chondrichthyes* (cartilaginous fishes), Superclass *Osteichthyes* (bony fishes), *Amphibia* (amphibians), *Reptilia* (=Sauropsida) (reptiles), *Aves* (birds), and *Mammalia* (mammals).

Chordates are a highly specialized and successful group of animals. Generally large in size, they occupy diverse habitats and have specialized niches and behaviors. All chordates are important in biological food chains and ecological communities.

Table 12.1 Representatives of the Phylum Chordata

Subphyla and Representative Kinds	Characteristics
Tunicata — tunicates	Marine, larvae are free-swimming and have notochord, gill slits, and dorsal hollow nerve cord; most adults are sessile (attached), filter-feeders, saclike animals
Cephalochordata — lancelets, amphioxus	Marine, segmented, elongated body with notochord extending the length of the body; cirri surrounding the mouth for obtaining food
Vertebrata — agnathans (lampreys and hagfishes), fishes (cartilaginous and bony), amphibians, reptiles, birds, mammals	Aquatic and terrestrial forms; distinct head and trunk supported by a series of cartilaginous or bony vertebrae in the adult; closed circulatory system and ventral heart; well-developed brain and sensory organs

Table 12.2 Representatives of the Subphylum Vertebrata

Taxa and Representative Kinds	Characteristics
Class Agnatha	Eel-like and aquatic; sucking mouth (some parasitic); lack jaws and paired appendages
Subclass Myxini — hagfishes	Terminal mouth with buccal funnel absent; nasal sac connected to pharynx; four pairs of tentacles; five to ten pairs pharyngeal pouches
Subclass Petromyzontida — lampreys	Suctorial mouth with rasping teeth; nasal sac not connected to buccal cavity; seven pairs of pharyngeal pouches
Infraphylum Gnathostomata	Jawed vertebrates; most with paired appendages
Class Chondrichthyes — sharks, rays, and skates	Cartilaginous skeleton; placoid scales; most have spiracle; spiral valve in digestive tract
Class Osteichthyes	Bony fishes; gills covered by bony operculum; most have swim bladder
Subclass Sarcopterygii	Bony skeleton; lobe-finned; paired pectoral and pelvic fins
Subclass Actinopterygii	Bony skeleton; most have dermal scales; ray-finned
Class Amphibia — salamanders, frogs, and toads	Larvae have gills and adults have lungs; scaleless skin (except apoda); an incomplete double circulation; three-chambered heart
Class Reptilia (= Sauropsida) — reptiles and birds*	Amniotic egg; epidermal scales; three- or four-chambered heart; lungs
Class Aves — birds*	Homeothermous (warm-blooded); feathers; toothless; air sacs; four-chambered heart with right aortic arch
Class Mammalia — mammals	Homeothermous; hair; mammary glands; most have seven cervical vertebrae; muscular diaphragm; three auditory ossicles; four-chambered heart with left aortic arch

* Birds and crocodilians are members of Archosauria, which include the dinosaurs. For convenience we treat them traditionally as a separate class.

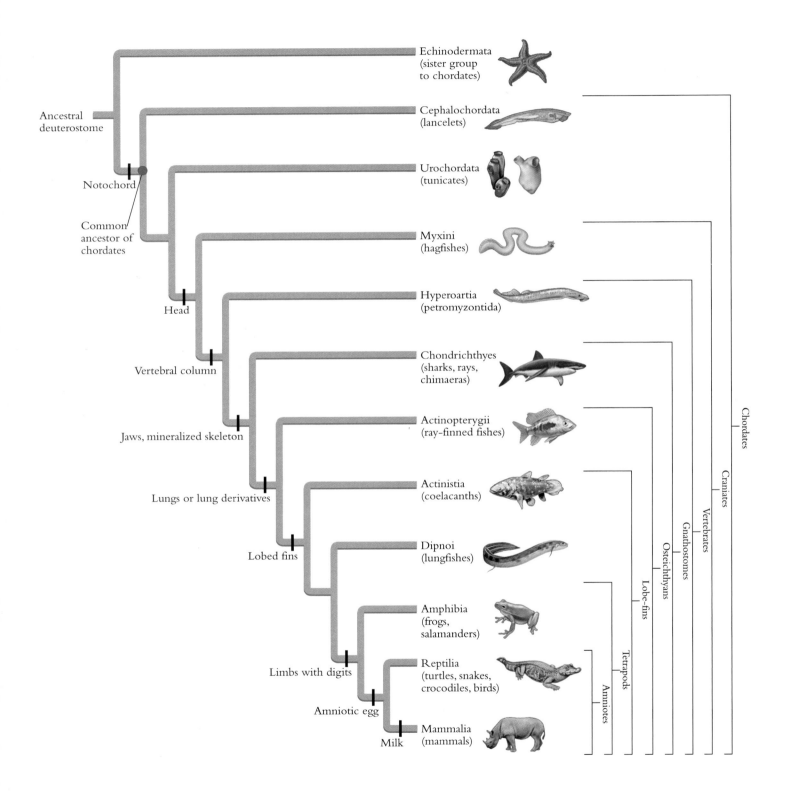

Figure 12.1 Phylogenetic relationships and classification of the major, extant, chordate lineages.

Phylum Chordata

Figure 12.2 Example chordates include: (a) an ink-spot sea squirt or tunicate, *Polycarpa aurata*, (b) a lancelet, *Branchiostoma* sp. (c) a hammer head shark, *Sphyrna tiburo*, (d) a giant grouper, *Epinephelus lanceolatus*, (e) a bullfrog, *Rana catesbeiana*, (f) a lazuli bunting, *Passerina amoena*, (g) a red squirrel, *Tamiasciurus hudsonicus*, and (h) a chimpanzee, *Pan troglodytes*.

Subphylum Tunicata

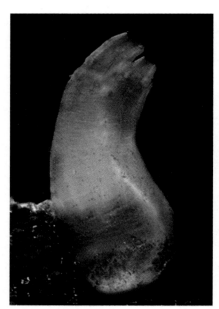

Figure 12.3 An adult tunicate, *Ciona intestinalis*.

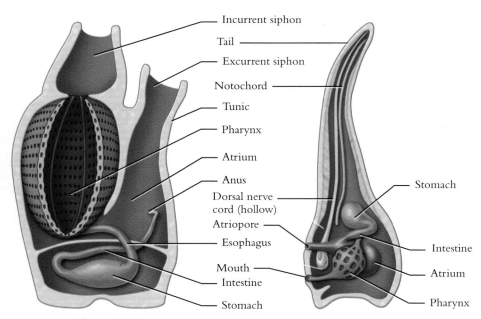

Figure 12.4 A diagram of a tunicate, (a) adult and (b) a larva.

Subphylum Cephalochordata

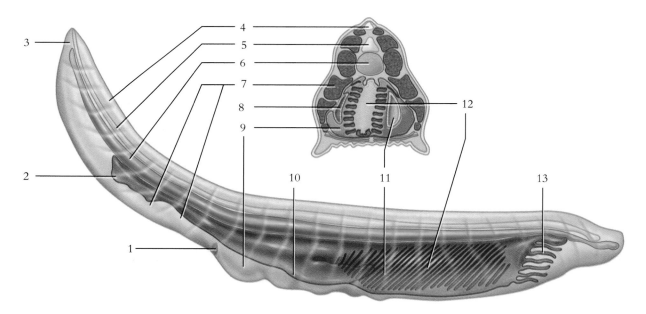

Figure 12.5 A diagram of a lancelet, *Branchiostoma* sp.
1. Atriopore
2. Anus
3. Caudal fin
4. Fin rays
5. Dorsal nerve cord
6. Notochord
7. Myomeres
8. Gonad
9. Atrium
10. Intestine
11. Digestive caecum
12. Pharynx with gill slits
13. Wheel organ

Chordata

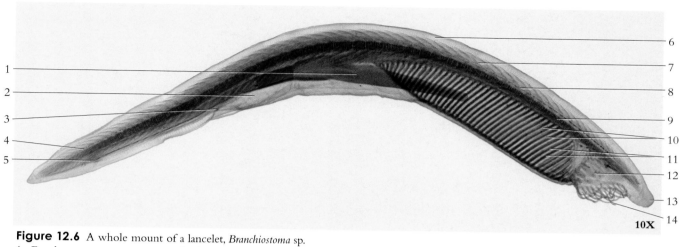

Figure 12.6 A whole mount of a lancelet, *Branchiostoma* sp.
1. Esophagus
2. Atrium
3. Atriopore
4. Caudal fin
5. Anus
6. Fin rays
7. Myomeres
8. Dorsal nerve cord
9. Notochord
10. Gill slits
11. Gill bars
12. Wheel organ
13. Rostrum
14. Oral cirri

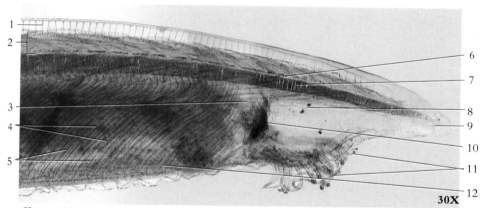

Figure 12.7 An anterior view of the anatomy of a lancelet, *Branchiostoma* sp.
1. Fin rays
2. Myomeres
3. Velum
4. Gill slits
5. Gill bars
6. Dorsal nerve cord
7. Notochord
8. Oral hood
9. Rostrum
10. Wheel organ
11. Oral cirri
12. Pharynx

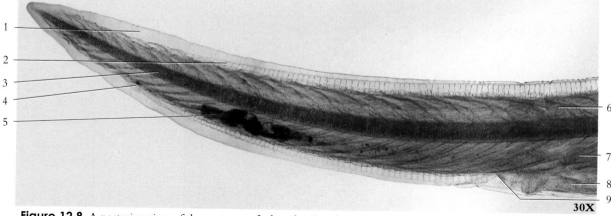

Figure 12.8 A posterior view of the anatomy of a lancelet, *Branchiostoma* sp.
1. Caudal fin
2. Fin rays
3. Notochord
4. Anus
5. Intestine
6. Myomeres
7. Midgut
8. Atrium
9. Atriopore

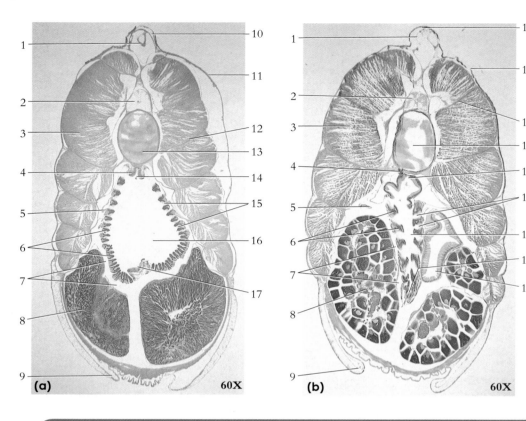

Figure 12.9 A transverse section through the pharyngeal region of (a) a male, and (b) a female lancelet, *Branchiostoma* sp.

1. Fin ray
2. Dorsal nerve cord
3. Myomere
4. Dorsal aorta
5. Nephridium
6. Gill bars
7. Atrium
8. Testis (male)
   Ovary (female)
9. Metapleural fold
10. Dorsal fin
11. Epidermis
12. Myoseptum
13. Notochord
14. Epibranchial groove
15. Gill slits
16. Pharynx
17. Endostyle (hypobranchial groove)
18. Hepatic caecum (liver)

Subclass Petromyzontida

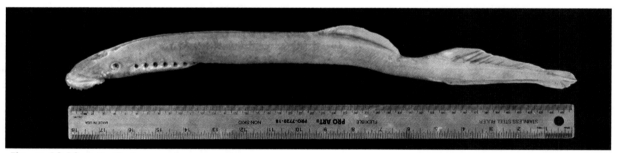

Figure 12.10 A Pacific lamprey, *Lampetra tridentata*.

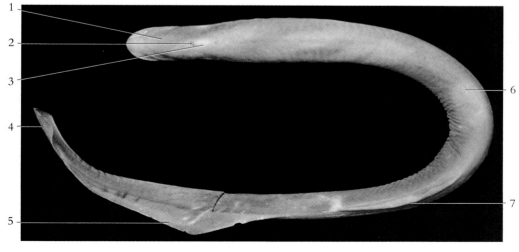

Figure 12.11 A dorsal view of the external anatomy of a marine lamprey, *Petromyzon marinus*.

1. Head
2. Nostril
3. Pineal body
4. Caudal fin
5. Posterior dorsal fin
6. Trunk
7. Anterior dorsal fin

Chordata

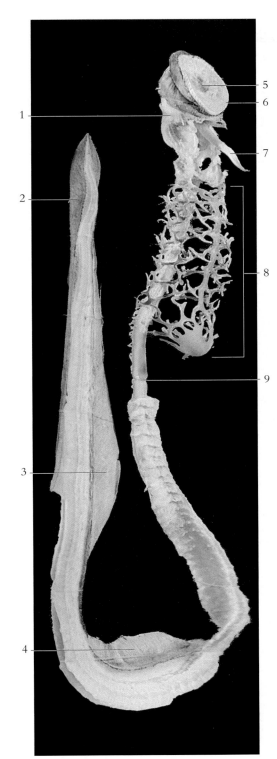

Figure 12.12 The cartilaginous skeleton of a marine lamprey.
1. Cranium
2. Caudal fin
3. Posterior dorsal fin
4. Anterior dorsal fin
5. Buccal cavity
6. Annular cartilage
7. Lingual cartilage
8. Branchial basket
9. Notochord

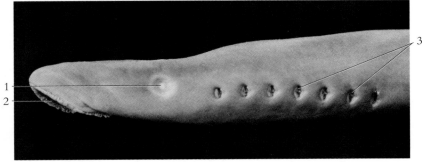

Figure 12.13 A lateral view of the anterior anatomy of a marine lamprey.
1. Eye
2. Buccal funnel
3. External gill slits

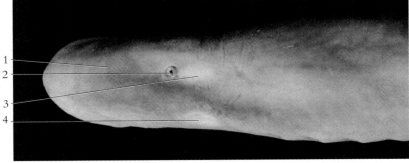

Figure 12.14 A dorsal view of the anterior anatomy of a marine lamprey.
1. Head
2. Nostril
3. Pineal body
2. Eye

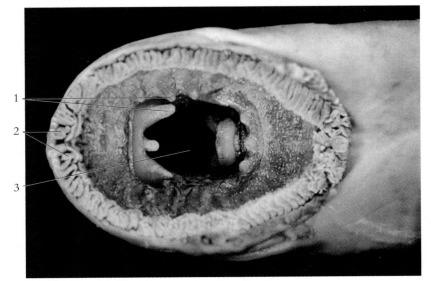

Figure 12.15 The oral region of a marine lamprey.
1. Horny teeth
2. Buccal papillae
3. Mouth

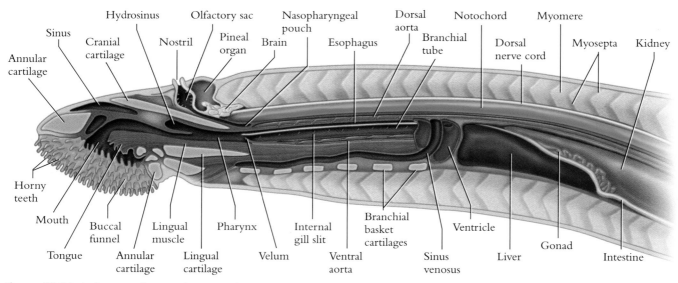

Figure 12.16 A diagram of a sagittal section of a marine lamprey.

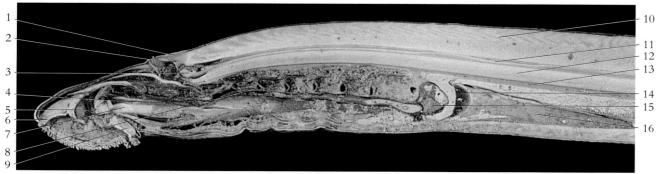

Figure 12.17 A sagittal section through the anterior region of a lamprey.

1. Pineal organ
2. Nostril
3. Brain
4. Pharynx
5. Mouth
6. Annular cartilage
7. Lingual cartilage
8. Internal gill slit
9. Buccal muscle
10. Myomeres
11. Dorsal nerve cord
12. Notochord
13. Dorsal aorta
14. Atrium
15. Ventricle
16. Liver

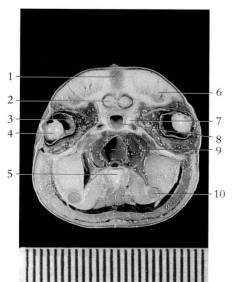

Figure 12.18 A transverse section through the head at the level of the eyes of a lamprey.

1. Pineal organ
2. Brain
3. Retina of eye
4. Lens of eye
5. Lingual cartilage
6. Myomere
7. Cranial cartilage
8. Nasopharyngeal pouch
9. Pharynx
10. Pharyngeal gland

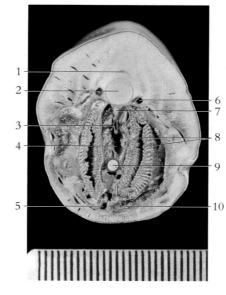

Figure 12.19 A transverse section through the body at the level of the fourth gill slit of a lamprey.

1. Dorsal nerve cord
2. Notochord
3. Esophagus
4. Branchial tube
5. Ventral jugular vein
6. Anterior cardinal vein
7. Dorsal aorta
8. Gill filaments
9. Ventral aorta
10. Branchial pouch

Chondrichthyes and Osteichthyes

Chapter 13

Fishes are the vertebrate organisms most highly adapted for an aquatic environment. Many of the specialized structures of fishes enable excellent exploitation of the dense water medium in which they are confined. Gills enable extraction of oxygen from the water. Buoyancy is maintained through physiological adaptations that may or may not involve swim bladders. Streamlined locomotion is achieved by body shapes, positions of fins, and placement of muscle masses. The sense organs of fishes are highly developed to respond to stimuli passing through a water medium.

Fishes include the organisms in three groups of vertebrates. *Agnatha* includes the jawless fishes (see Chapter 12), such as the lampreys and hagfishes. *Chondrichthyes* includes the cartilaginous fishes, such as the skates, rays, and sharks. The bony fishes, Osteichthyes, include the ancient lobe-fin fishes *Sarcopterygii* and the more modern ray-fin fishes *Actinopterygii*. The cartilaginous fishes and the bony fishes are described and depicted in this chapter.

Chondrichthyes (Cartilaginous Fishes)

Nearly 850 species of skates, rays, sharks, and chimaeras are contained within the class Chondrichthyes. The subclass *Elasmobranchii* contains the skates, rays, and sharks, which number 800 species. The subclass *Holocephali* contains the 25 species of chimaeras. Chimaeras are remnants of ancient elasmobranchs that were much more common during the Devonian period nearly 300 million years ago.

Some of the characteristics of Chondrichthyes include:

1. *Fusiform body* with a *heterocercal tail*. The body of most sharks is tapered at both ends—like the fuselage of an airplane. This streamlined shape is ideal for rapid swimming. The caudal vertebrae turn upward, forming the heterocercal tail. Paired pectoral and pelvic fins are controlled by muscles attached to the girdles (cartilage that provides structural support and leverage). Prominent dorsal fins provide stability while swimming.
2. Ventral-positioned mouth and a *spiracle* exiting from the oral cavity. Well-developed jaws surround the mouth and support many *homodont teeth* (similarly shaped teeth). Most Chondrichthyes are excellent predators.
3. *Placoid scales* cover the skin. Composed of enamel and dentin, the placoid scales provide excellent protection to these animals. The teeth of Chondrichthyes are modified placoid scales that are continuously replaced as they are shed.
4. *Cartilaginous endoskeleton.* Although derived from early Devonian ancestors that had bony endoskeletons, Chondrichthyes maintain a cartilaginous endoskeleton even as adult animals.
5. Five to seven pairs of *gills*. Except in chimaeras, the gills of Chondrichthyes are not covered by an operculum.
6. Two-chambered heart with several pairs of aortic arches and a dorsal and ventral aorta. High concentrations of urea and trimethylamine oxide are maintained within the blood and aid buoyancy. Chondrichthyes lack swim bladders and lungs.
7. *Spiral valve* in digestive system. The digestive system has distinct portions for specialized functions. The spiral valve delays the passage of food and increases the absorptive surface. Attached to the rectum is the *rectal gland,* which secretes a fluid containing sodium chloride that assists the kidney in regulating the salt concentration of the blood.
8. Sensory portions of brain are well-developed. Chondrichthyes have a specialized *lateral line system* for detection of water vibrations. The lateral line system connects to the three pairs of semicircular canals. Olfaction is well-developed, but vision is considered poor.
9. Males have *claspers* and fertilization is internal, with *oviparous, ovoviviparous,* or *viviparous* development. Cartilaginous fishes have *mesonephros*–type kidneys.

Osteichthyes (Bony Fishes)

Superclass Osteichthyes includes the bony fishes, of which there are approximately 24,600 species within the class *Sarcopterygii* (lobe-fin fishes) and class Actinopterygii (ray-fin fishes). There are few living species of lobe-fin fishes, but the group is highly significant because from ancient extinct members of these fishes descended the amphibians and indeed all tetrapod vertebrates. Lobe-fin fishes have bony elements supporting their paired pectoral and pelvic fins. Lungfishes within the group respire through gills as well as lungs. They also have nostrils that open into the mouth.

Some authorities divide the lobe-finned fishes into three subclasses. In this scheme, subclass Coelacanthimorpha is the designation of the lobe-finned fishes containing one living order and one genus, *Latimeria*. Subclass Dipnoi contains two living orders of lungfishes and three genera: *Neoceratodus, Lepidosiren,* and *Protopterus*. The third subclass is the Tetrapodomorpha, known only from fossils, which possess a combination of fish and tetrapod characteristics. All of the remaining bony fishes belong to the class Actinopterygii (ray-finned fish).

Some of the characteristics of actinopterygians (ray–finned fishes) include:

1. Most have *fusiform body* with *homocercal tail* (uniform arrangement of supporting rays).
2. *Dermal scales* present within glandular skin. The three kinds of scales are *ganoid, cycloid,* and *ctenoid*. A few bony fishes lack scales. *Mucous glands* are abundant within the skin and secrete a protective mucous film.

3. Mouth terminal. The mouth has well-developed jaws, and most fishes have numerous homodont teeth. Paired olfactory sacs may or may not open into the mouth.
4. Bony *operculum*. The operculum protects the fleshy gills that are supported by gill arches.
5. *Swim bladder.* The swim bladder is a hydrostatic organ that may or may not connect to the pharynx. Some fishes have a direct connection of the pharynx to the swim bladder through a *pneumatic duct*. Other fishes lack a connection and have instead a *gas gland* (or *red body*) that extracts gases from the blood to inflate the swim bladder.
6. Two-chambered heart with four pairs of aortic arches. The blood of fishes contains nucleated red blood cells.
7. Sensory portions of the brain are well-developed. The optic lobes are large; there are three pairs of semicircular canals and 10 pairs of cranial nerves.
8. Separate sexes. Fertilization is usually external. Most bony fishes are oviparous, but some species are ovoviviparous or viviparous.

Class Chondrichthyes

Figure 13.1 A black tip reef shark, *Carcharhinus melanopterus*.

Figure 13.2 A gray reef shark, *Carcharhinus amblyrhynchos*.

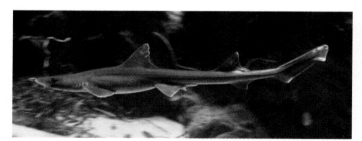

Figure 13.3 A gray smoothhound shark, *Mustelus californicus*.

Figure 13.4 A guitarfish, *Rhina ancylostoma*.

Figure 13.5 A blue-spotted stingray, *Taeniura lymma*.

Figure 13.6 A chimaera, *Hydrolagus colliei*.

Chondrichthyes and Osteichthyes

Figure 13.7 A leopard shark, *Triakis semifasciata*.

Figure 13.8 A round stingray, *Urobatis halleri*.

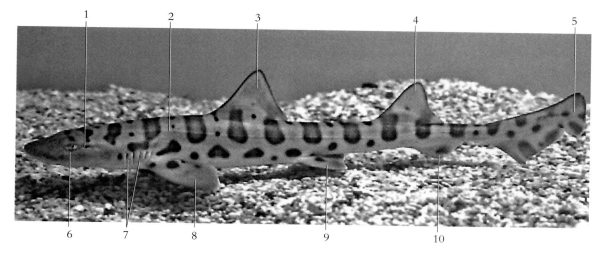

Figure 13.9 A lateral view of the leopard shark, *Triakis semifasciata*.
1. Spiracle
2. Lateral line
3. Anterior dorsal fin
4. Posterior dorsal fin
5. Caudal fin (heterocercal tail)
6. Eye
7. Gill slits
8. Pectoral fin
9. Pelvic fin
10. Anal fin

Figure 13.10 A photomicrograph of placoid scales.

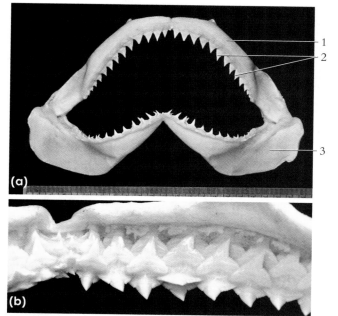

Figure 13.11 Shark jaws (a) and (b) a detailed view showing rows of replacement teeth (scale in mm).
1. Palatopterygoquadrate cartilage (upper jaw)
2. Placoid teeth
3. Meckel's cartilage (lower jaw)

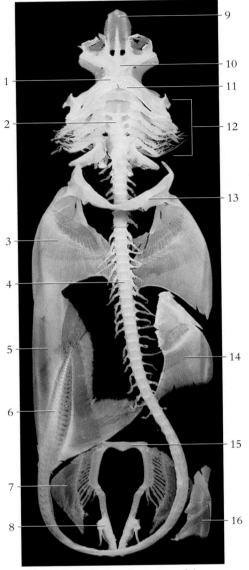

Figure 13.12 A ventral view of the cartilaginous skeleton of a male dogfish shark.

1. Palatopterygoquadrate cartilage (upper jaw)
2. Hypobranchial cartilage
3. Pectoral fin
4. Trunk vertebrae
5. Caudal fin
6. Caudal vertebrae
7. Pelvic fin
8. Clasper
9. Rostrum
10. Chondrocranium
11. Meckel's cartilage (lower jaw)
12. Visceral arches (gill arches)
13. Pectoral girdle
14. Anterior dorsal fin
15. Pelvic girdle
16. Posterior dorsal fin

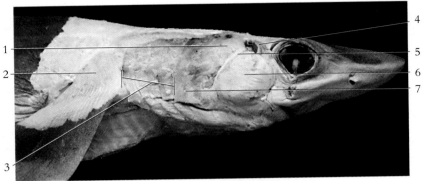

Figure 13.13 The musculature of the jaw, gills, and pectoral fin of a dogfish shark.

1. 2nd dorsal constrictor
2. Levator of pectoral fin
3. 3rd through 6th ventral constrictors
4. Spiracular muscle
5. Facial nerve (hyomandibular branch)
6. Mandibular adductor
7. 2nd ventral constrictor

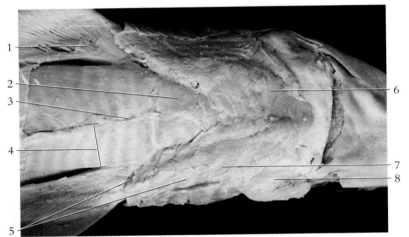

Figure 13.14 A ventral view of the hypobranchial musculature of the dogfish shark.

1. Depressor of pectoral fin
2. Common coracoarcual
3. Linea alba
4. Hypaxial muscle
5. 3rd through 6th ventral constrictors
6. 1st ventral constrictor
7. 2nd ventral constrictor
8. Mandibular adductor

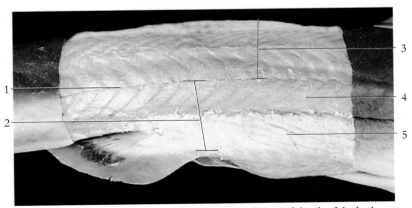

Figure 13.15 A lateral view of the axial musculature of the dogfish shark.

1. Horizontal septum
2. Hypaxial myotome portion
3. Epaxial myotome portion
4. Lateral bundle of myotomes
5. Ventral bundle of myotomes

Chondrichthyes and Osteichthyes

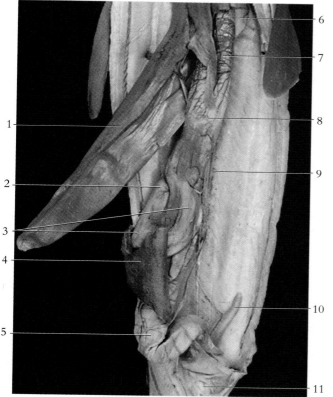

Figure 13.16 The internal anatomy of a male dogfish shark.
1. Right lobe of liver (reflected)
2. Pyloric sphincter valve
3. Stomach (pyloric region)
4. Spleen
5. Ileum (intestine)
6. Testis
7. Esophagus
8. Stomach (cardiac region)
9. Kidney
10. Rectal gland
11. Cloaca

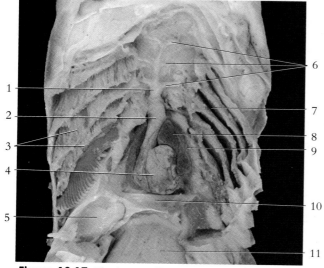

Figure 13.17 The heart, gills, and associated vessels of a dogfish shark.
1. Ventral aorta
2. Conus arteriosus
3. Gills
4. Ventricle
5. Pectoral girdle (cut)
6. Afferent branchial arteries
7. Gill cleft
8. Atrium
9. Pericardial cavity
10. Transverse septum
11. Liver

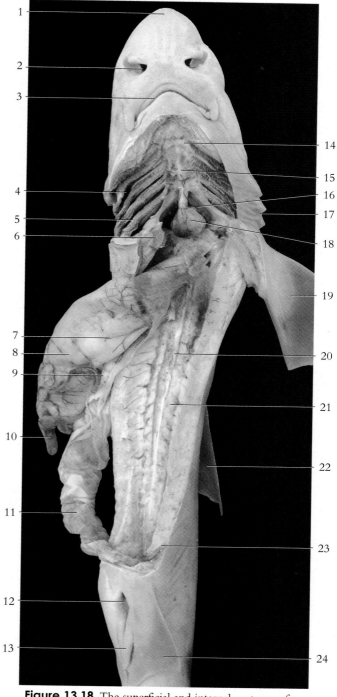

Figure 13.18 The superficial and internal anatomy of a dogfish shark (liver removed).
1. Rostrum
2. Nostril
3. Mouth
4. Gill cleft
5. Gill
6. Pectoral girdle (cut)
7. Gastrosplenic artery
8. Stomach (reflected)
9. Pancreas
10. Spleen (reflected)
11. Intestine (reflected)
12. Cloaca
13. Clasper
14. Afferent branchial artery
15. Ventral aorta
16. Atrium
17. Gill slit
18. Ventricle of heart
19. Pectoral fin
20. Dorsal aorta
21. Kidney
22. Dorsal fin (anterior)
23. Rectal gland
24. Pelvic fin

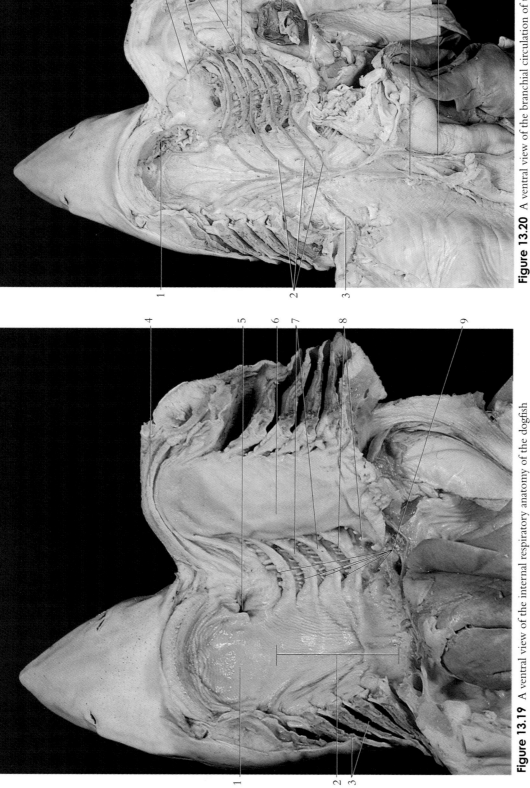

Figure 13.19 A ventral view of the internal respiratory anatomy of the dogfish shark (lower jaw cut and reflected).

1. Oral cavity
2. Pharynx
3. Parabranchial chambers
4. Teeth
5. Spiracle
6. Tongue
7. Gill arches
8. Gill rakers
9. Internal gill slits (5)

Figure 13.20 A ventral view of the branchial circulation of the dogfish shark (lower jaw cut and reflected).

1. Stapedial artery
2. Efferent branchial arteries
3. Subclavian artery
4. External carotid artery
5. Afferent branchial arteries
6. Ventral aorta
7. Hypobranchial artery
8. Heart (atrium)
9. Anterior epigastric artery
10. Dorsal aorta
11. Celiac artery

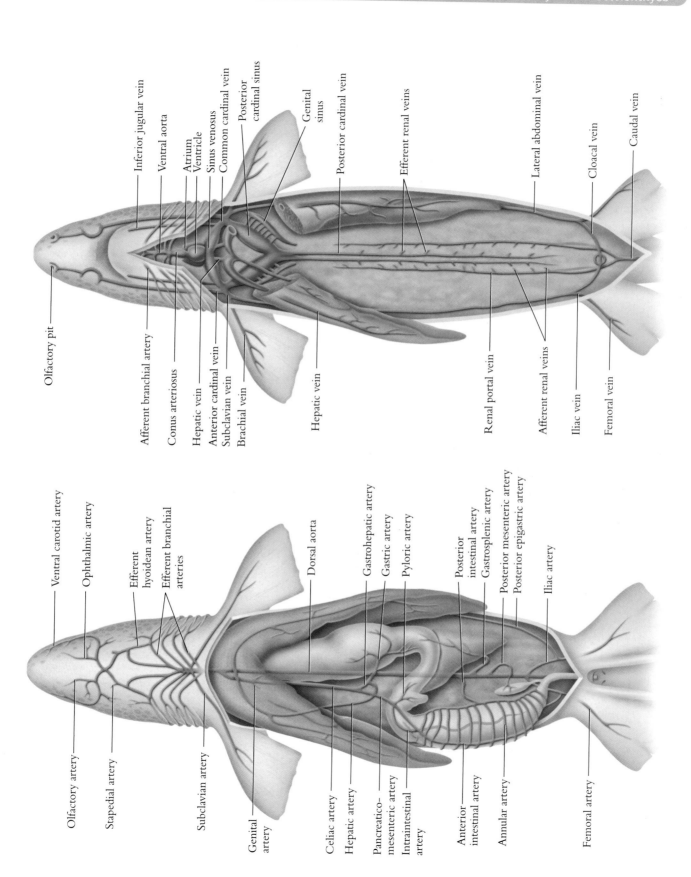

Figure 13.22 The veins of a dogfish shark.

Figure 13.21 The arteries of a dogfish shark.

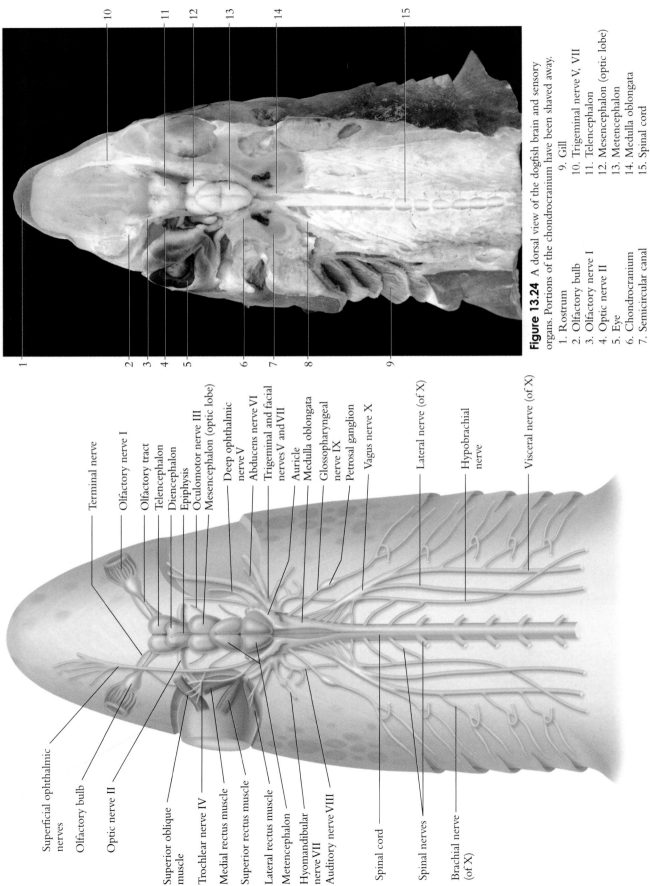

Figure 13.24 A dorsal view of the dogfish brain and sensory organs. Portions of the chondrocranium have been shaved away.

1. Rostrum
2. Olfactory bulb
3. Olfactory nerve I
4. Optic nerve II
5. Eye
6. Chondrocranium
7. Semicircular canal
8. Vagus nerve X
9. Gill
10. Trigeminal nerve V, VII
11. Telencephalon
12. Mesencephalon (optic lobe)
13. Metencephalon
14. Medulla oblongata
15. Spinal cord

Figure 13.23 A dorsal view of the dogfish brain, cranial nerves, and eye muscles.

Class Osteichthyes - Subclass Sarcopterygii

Figure 13.25 The coelacanth, *Latimeria chalumnae*, a lobe-fin fish, was once thought to be extinct.

Figure 13.26 The African lungfish, *Neoceratodus forsteri*.

Class Osteichthyes - Subclass Actinopterygii

Figure 13.27 Pacific sardines, *Sardinops sagax*.

Figure 13.28 A lake trout, *Salvelinus namaycush*.

Figure 13.29 A lookdown fish, *Seiene vomer*.

Figure 13.30 A red discus fish, *Symphysodon aequifasciatus*.

Figure 13.31 A frogfish, *Antennarius* sp.

Figure 13.32 Jack mackerel, *Trachurus declivis*.

Figure 13.33 A tomato clownfish, *Amphiprion melanopus*.

Figure 13.34 Chum salmon, *Oncorhynchus keta*.

Figure 13.35 A lionfish, *Pterois* sp.

Figure 13.36 The external structures of a grouper, *Mycteroperca bonaci*.

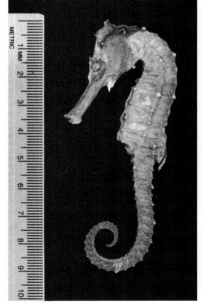

Figure 13.37 A prepared specimen of a seahorse.

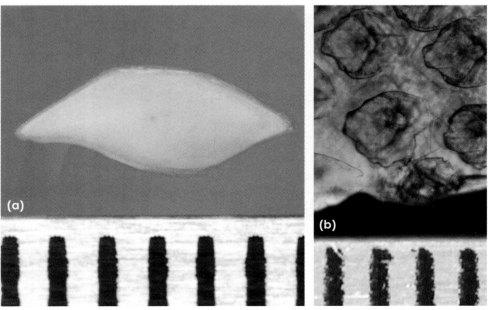

Figure 13.38 Ganoid scales, present in primitive fishes like the gar, are composed of silvery ganoin on the top surface and bone on the bottom. Two different sizes are shown (a) and (b) (scale in mm).

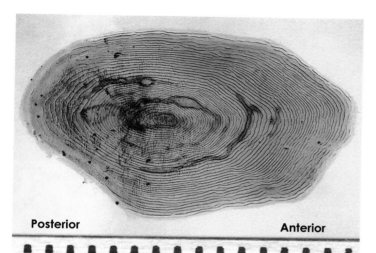

Figure 13.39 Cycloid scales, along with ctenoid scales (fig. 13.40), are found on advanced bony fishes. They are much thinner and more flexible than ganoid scales and overlap each other (scale in mm).

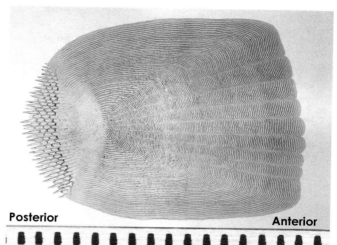

Figure 13.40 Ctenoid scales differ from cycloid scales in that they have comblike ridges on the exposed edge. This is thought to improve swimming efficiency (scale in mm).

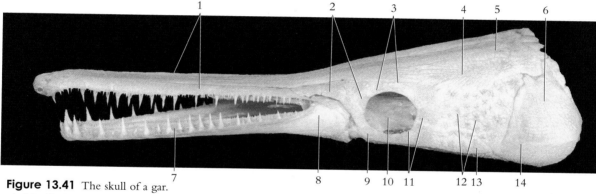

Figure 13.41 The skull of a gar.
1. Maxillaries
2. Preorbitals
3. Supraorbitals
4. Pterotic
5. Parietal
6. Opercular
7. Dentary
8. Angular
9. Infraorbital
10. Orbit
11. Postorbitals
12. Cheek plates
13. Preopercular
14. Subopercular

Figure 13.42 The skeleton of a perch.
1. Anterior dorsal fin
2. Fin spines
3. Neurocranium
4. Premaxilla
5. Maxilla
6. Dentary
7. Opercular bones
8. Pectoral fin
9. Pelvic fin
10. Vertebral column
11. Posterior dorsal fin
12. Soft rays
13. Caudal fin
14. Neural spine
15. Haemal spine
16. Anal fin
17. Ribs

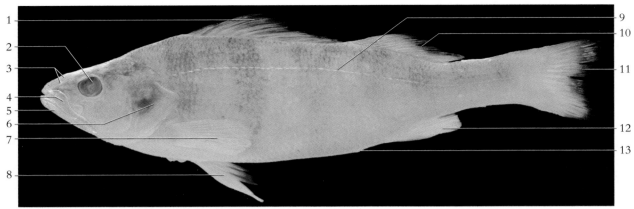

Figure 13.43 The external anatomy of a perch.
1. Anterior dorsal fin
2. Eye
3. Nostrils
4. Mandible
5. Dentary
6. Operculum
7. Pectoral fin
8. Pelvic fin
9. Lateral line
10. Posterior dorsal fin
11. Caudal fin
12. Anal fin
13. Anus

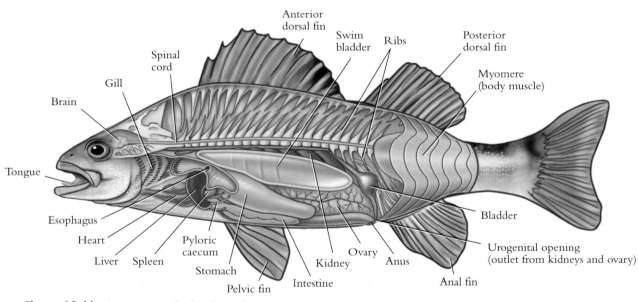

Figure 13.44 The anatomy of a female perch.

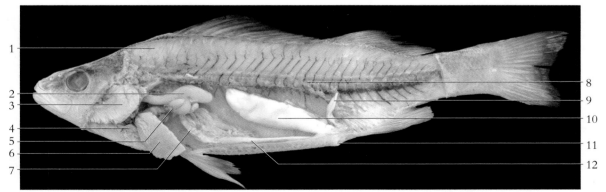

Figure 13.45 The viscera of a perch.
1. Epaxial muscles
2. Stomach
3. Gill
4. Heart
5. Pyloric caecum
6. Liver (cut)
7. Pancreas
8. Vertebrae
9. Urinary bladder
10. Gonad
11. Anus
12. Intestine

Amphibia

Chapter 14

There are an estimated 7,690 species of amphibians contained within the class Amphibia. Included in this group are caecilians, salamanders, frogs, and toads. Amphibians are vertebrate animals that are transitional in body structure and behavior between aquatic and terrestrial environments. Most larval amphibians are adapted for an aquatic environment. They have gills, elongated tails for swimming, a lateral line system, and a digestive tract adapted for utilization of plant material. Adult amphibians generally occupy a moist terrestrial habitat. Adults breathe by lungs (some have gills), are thin-skinned, and have a specialized three-chambered heart. The amphibian egg is deposited in the water and is fishlike in that it lacks a shell and an amnion.

There are three orders of living amphibians. Order Gymnophiona includes the estimated 200 species of caecilians. These amphibians lack pectoral and pelvic girdles and appendages, and their tails are short or absent. Most caecilians have *mesodermal* scales. Order Caudata includes the estimated 550 species of salamanders. Salamanders have distinct body regions (head, trunk, and tail) and usually two paired appendages. Scales are lacking in salamanders. Order Anura includes the estimated 4,800 species of frogs and toads. These amphibians have smooth skin (lack scales), fused head and trunk, and two pairs of limbs with the hindlimbs adapted for jumping and swimming.

Some of the characteristics of amphibians include:

1. *Variable body forms*—Some amphibians lack limbs; others have limbs adapted for a generalized terrestrial locomotion. Others have highly specialized hindlimbs for jumping and swimming. In many, webbed feet are present, with no true claws.

2. *Smooth glandular skin*—The skin glands in many species produce poison. Pigment producing cells, called chromatophores, are common in many amphibians.

3. *Large mouth, small homodont teeth, and many with a protrusile tongue*—All amphibians have paired nostrils opening into the oral cavity.

4. *Respiration through a variety of means*—The moistened skin in many functions as a respiratory surface. Lungs are present in all but a few salamanders. External gills are present in larvae and may persist into adult forms of certain salamanders.

5. *Three-chambered heart*—The amphibian heart has two atria and one ventricle. Blood circulation is through a pulmo-cutaneous circuitry to the lungs and skin and then back to the heart to be pumped to the body.

6. *Brain is generalized*—Although the sensory portions of the brain may be well-developed in some, most amphibians have a simplified brain. Like the fishes, amphibians have ten pairs of cranial nerves.

7. *Paired opisthonephric kidneys and a urinary bladder.*

8. *Separate sexes*—Fertilization is internal by spermatophores in most caecilians and salamanders and external in most frogs and toads. Most amphibians are oviparous, some are ovoviviparous, and some are viviparous.

Amphibians are extremely important in semiaquatic communities and the general food chain. Most amphibians feed upon insects, and they themselves are a source of food for many fishes, birds, and mammals.

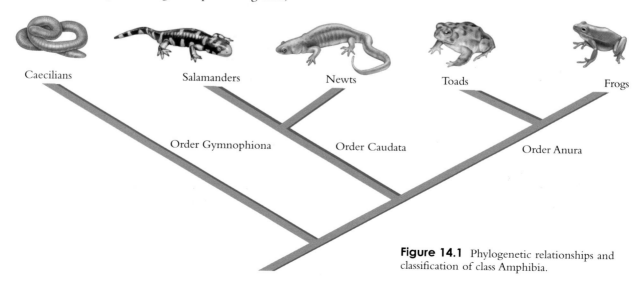

Figure 14.1 Phylogenetic relationships and classification of class Amphibia.

Class Amphibia

Figure 14.2 Examples of class Amphibia include: (a) a Cameroon caecilian, *Crotaphatrema bornmuelleri*, (b) an amphiuma, *Amphiuma means*, (c) a lesser siren, *Siren intermedia*, (d) an axolotl, *Ambystoma mexicanum*, (e) a tiger salamander, *Ambystoma tigrinum*, (f) a red mud salamander, *Pseudotriton ruber*, (g) an eastern newt, *Notophthalmus viridescens*, (h) a Woodhouse's toad, *Bufo woodhousii*, (i) a Colorado River toad, *Bufo alvarius*, (j) a blue-webbed flying treefrog, *Rhacophorus nigropalmatus*, (k) a canyon tree frog, *Hyla arenicolor*, and (l) a red-eyed tree frog, *Agalychnis callidryas*.

Figure 14.3 Cameroon caecilian, *Crotaphatrema bornmuelleri*. The rings or annuli can clearly be seen. These give caecilians an earthworm-like appearance.

Figure 14.4 An amphiuma, *Amphiuma means*. Note the small vestigial leg. The light colored dots are a lateral line system that aids in hunting.
1. Lateral line system
2. Vestigial limb

Figure 14.5 An axolotl, *Ambystoma mexicanum*. This individual is leucistic, a condition characterized by reduced pigmentation.

Figure 14.6 The marine or cane toad, *Bufo marinus*, is an introduced species to Hawaii and has caused many problems for native species.

Figure 14.7 The surface anatomy and body regions of the leopard frog, *Rana pipiens*.
1. Ankle
2. Knee
3. Foot
4. Eyes
5. Nostril
6. Tympanic membrane
7. Brachium
8. Antebrachium
9. Digits

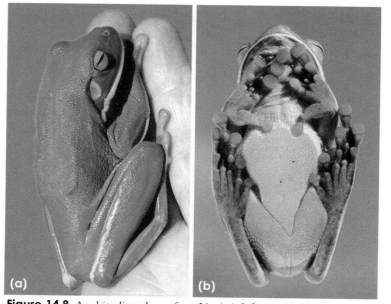

Figure 14.8 A white-lipped tree frog, *Litoria infrafrenata*. (a) The frog is crouched on a person's fingers. (b) The adhesive toe disks can be seen in a ventral view.

Order Caudata

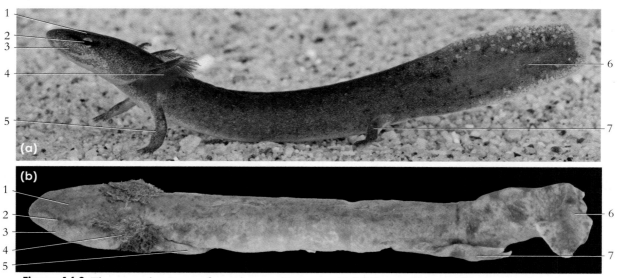

Figure 14.9 The external anatomy of a mud puppy, *Necturus* sp. (a) A living specimen and (b) a preserved specimen.
1. Cranium   2. Eye   3. Dentary   4. External gill   5. Forelimb   6. Tail   7. Hind limb

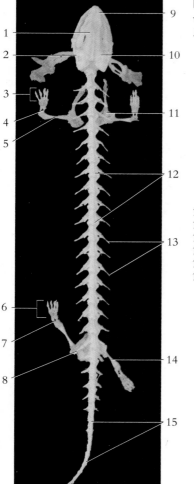

Figure 14.10 A dorsal view of the mud puppy, *Necturus*, skeleton.
1. Frontal bone
2. Squamosal bone
3. Manus (carpal bones, metacarpal bones, phalanges)
4. Ulna
5. Humerus
6. Pes (tarsal bones, metatarsal bones, phalanges)
7. Fibula
8. Pelvic girdle
9. Vomer bone
10. Parietal bone
11. Pectoral girdle
12. Trunk vertebrae
13. Ribs
14. Femur
15. Caudal vertebrae

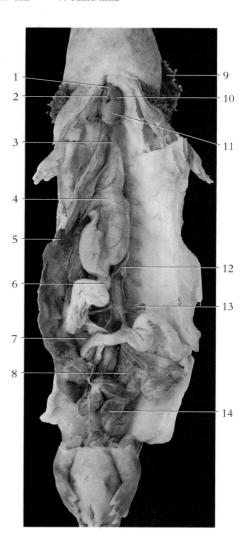

Figure 14.11 The internal anatomy of a mud puppy, *Necturus*.
1. Bulbus arteriosus
2. Conus arteriosus
3. Esophagus
4. Stomach
5. Liver
6. Duodenum
7. Gallbladder
8. Mesentery
9. External gills
10. Left atrium of heart
11. Ventricle of heart
12. Left lung
13. Pancreas
14. Colon

Order Anura

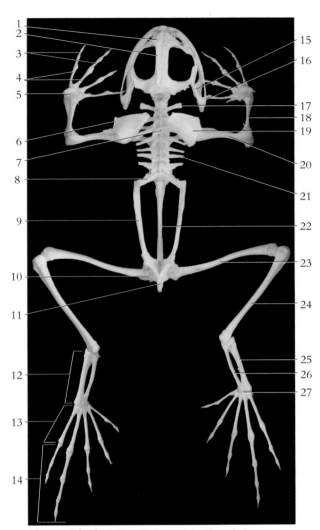

Figure 14.12 A dorsal view of the frog skeleton.
1. Nasal bone
2. Frontoparietal bone
3. Phalanges of digits
4. Metacarpal bones
5. Carpal bones
6. Scapula
7. Vertebra
8. Transverse process of sacral (9) vertebra
9. Ilium
10. Acetabulum
11. Ischium
12. Tarsal bones
13. Metatarsal bones
14. Phalanges of digits
15. Squamosal bone
16. Quadratojugal bone
17. Transverse process
18. Radioulna
19. Suprascapula
20. Humerus
21. Transverse process
22. Urostyle
23. Femur
24. Tibiofibula
25. Fibulare (calcaneum)
26. Tibiale (astragalus)
27. Distal tarsal bones

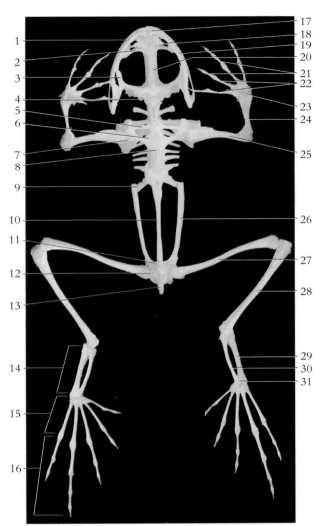

Figure 14.13 A ventral view of the frog skeleton.
1. Maxilla
2. Palatine
3. Pterygoid bone
4. Exoccipital bone
5. Clavicle
6. Coracoid
7. Glenoid fossa
8. Sternum
9. Transverse process of sacral (9th) vertebra
10. Urostyle
11. Pubis
12. Acetabulum
13. Ischium
14. Tarsal bones
15. Metatarsal bones
16. Phalanges of digits
17. Premaxilla
18. Vomer
19. Dentary
20. Parasphenoid bone
21. Phalanges of digits
22. Metacarpal bone
23. Carpal bones
24. Radioulna
25. Humerus
26. Ilium
27. Femur
28. Tibiofibula
29. Fibulare (calcaneum)
30. Tibiale (astragalus)
31. Distal tarsal bones

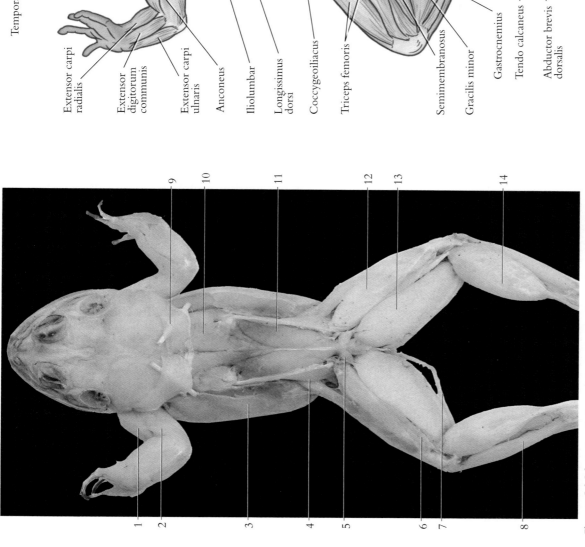

Figure 14.14 A dorsal view of the frog musculature (m. = muscle).
1. Deltoid m.
2. Anconeus m.
3. External abdominal oblique m.
4. Gluteus m.
5. Piriformis m.
6. Biceps femoris m.
7. Gracilis minor m.
8. Peroneus m.
9. Latissimus dorsi m.
10. Longissimus dorsi m.
11. Coccygeoiliacus m.
12. Triceps femoris m.
13. Semimembranosus m.
14. Gastrocnemius m.

Figure 14.15 A diagram of the dorsal frog musculature.

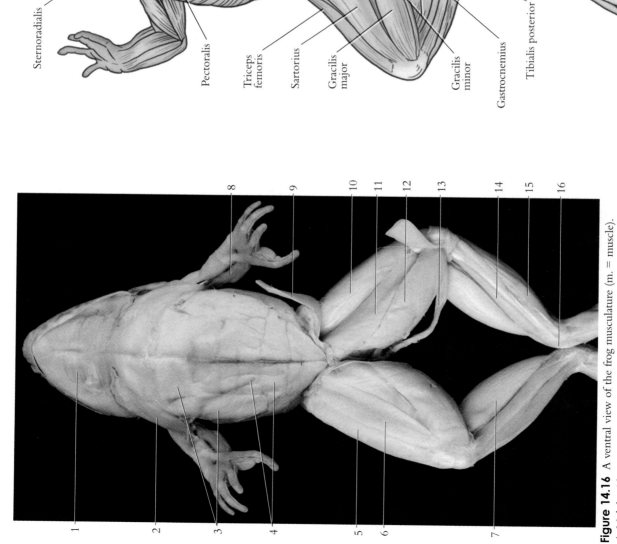

Figure 14.16 A ventral view of the frog musculature (m. = muscle).

1. Mylohyoid m.
2. Deltoid m.
3. Pectoralis m.
4. Rectus abdominis m.
5. Triceps femoris m.
6. Sartorius m.
7. Gastrocnemius m.
8. Palmaris longus m.
9. Sartorius m. (cut and reflected)
10. Triceps femoris m.
11. Adductor magnus m.
12. Gracilis major m.
13. Gracilis minor m.
14. Tibialis posterior m.
15. Tibialis anterior m.
16. Tendo calcaneus

Figure 14.17 A diagram of the ventral frog musculature.

Figure 14.18 A dorsal view of the leg muscles of a frog (m. = muscle).

1. Gluteus m.
2. Triceps femoris m.
3. Piriformis m.
4. Semimembranosus m. (cut and reflected)
5. Gracilis minor m.
6. Peroneus m.
7. Coccygeoiliacus m.
8. Triceps femoris m. (cut and reflected)
9. Iliacus internus m.
10. Biceps femoris m.
11. Adductor magnus m.
12. Semitendinosus m. (cut and reflected)
13. Gastrocnemius m.

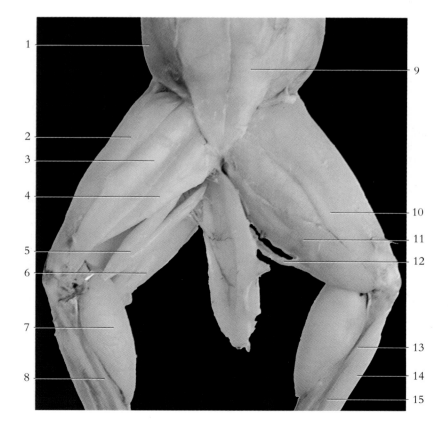

Figure 14.19 A ventral view of the leg muscles of a frog (m. = muscle).

1. External abdominal oblique m.
2. Triceps femoris m.
3. Sartorius m.
4. Adductor magnus m.
5. Semitendinosus m. (cut)
6. Semimembranosus m.
7. Gastrocnemius m.
8. Tibialis posterior m.
9. Rectus abdominis m.
10. Sartorius m.
11. Gracilis major m.
12. Gracilis minor m.
13. Extensor cruris m.
14. Tibialis anterior longus m.
15. Tibialis anterior brevis m.

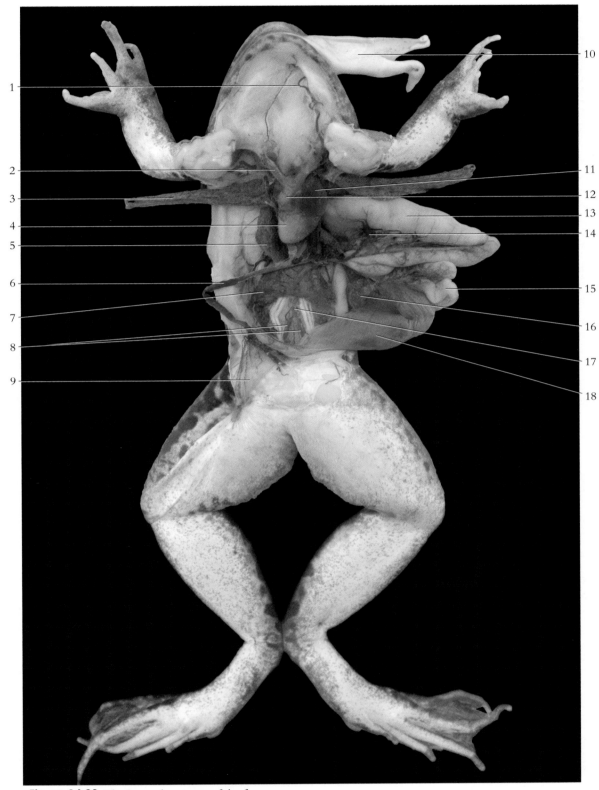

Figure 14.20 The internal anatomy of the frog.
1. External carotid artery
2. Truncus arteriosus
3. Lung (reflected)
4. Ventricle of heart
5. Liver (cut)
6. Ventral abdominal vein
7. Kidney
8. Iliac arteries
9. Bladder (reflected)
10. Tongue
11. Right atrium of heart
12. Conus arteriosus
13. Stomach (reflected)
14. Gastric vein
15. Small intestine
16. Spleen
17. Dorsal aorta
18. Large intestine (reflected)

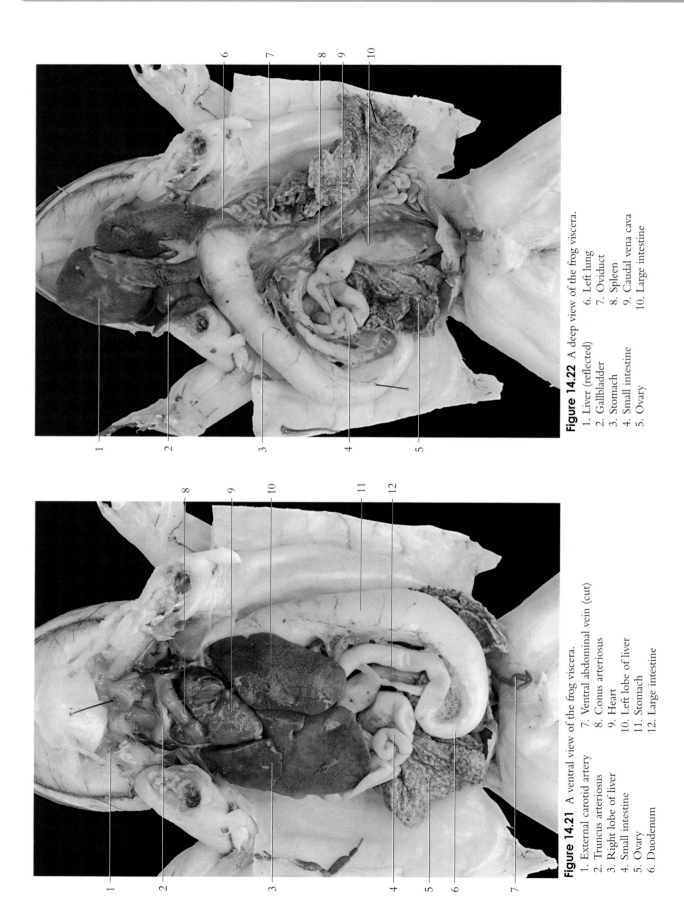

Figure 14.22 A deep view of the frog viscera.

1. Liver (reflected)
2. Gallbladder
3. Stomach
4. Small intestine
5. Ovary
6. Left lung
7. Oviduct
8. Spleen
9. Caudal vena cava
10. Large intestine

Figure 14.21 A ventral view of the frog viscera.

1. External carotid artery
2. Truncus arteriosus
3. Right lobe of liver
4. Small intestine
5. Ovary
6. Duodenum
7. Ventral abdominal vein (cut)
8. Conus arteriosus
9. Heart
10. Left lobe of liver
11. Stomach
12. Large intestine

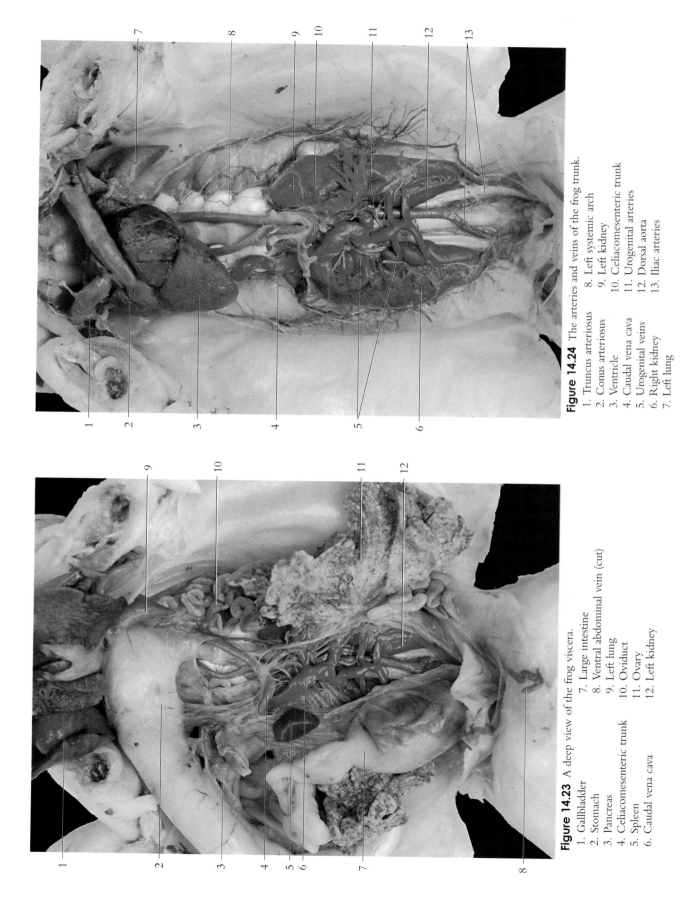

Figure 14.23 A deep view of the frog viscera.
1. Gallbladder
2. Stomach
3. Pancreas
4. Celiacomesenteric trunk
5. Spleen
6. Caudal vena cava
7. Large intestine
8. Ventral abdominal vein (cut)
9. Left lung
10. Oviduct
11. Ovary
12. Left kidney

Figure 14.24 The arteries and veins of the frog trunk.
1. Truncus arteriosus
2. Conus arteriosus
3. Ventricle
4. Caudal vena cava
5. Urogenital veins
6. Right kidney
7. Left lung
8. Left systemic arch
9. Left kidney
10. Celiacomesenteric trunk
11. Urogenital arteries
12. Dorsal aorta
13. Iliac arteries

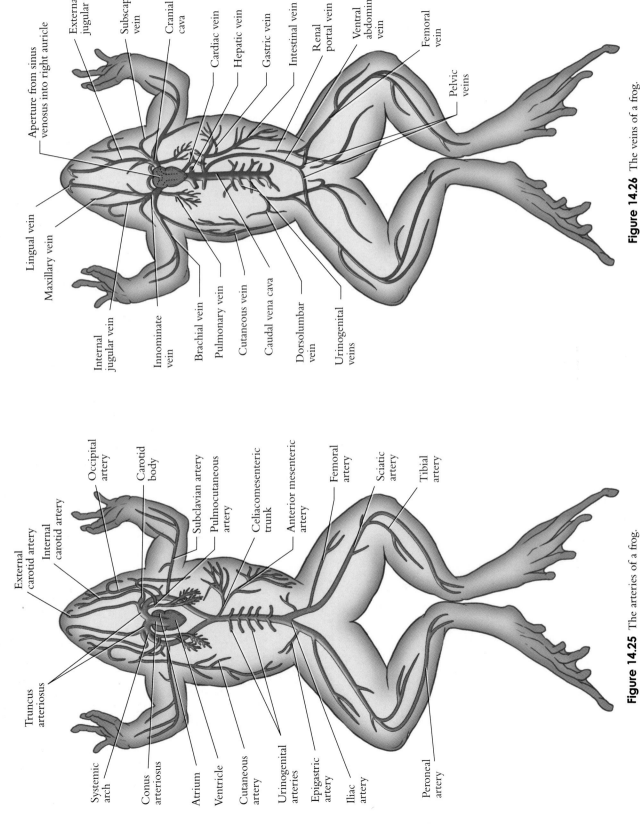

Figure 14.26 The veins of a frog.

Figure 14.25 The arteries of a frog.

Reptilia (= Sauropsida)

Chapter 15

Sauropsids are chordate, tetrapod amniotes that include several independent evolutionary lineages commonly referred to as reptiles (turtles, tuataras, lizards, snakes, crocodiles, and alligators). Sauropsids also include birds and their fossil ancestors, the dinosaurs (see fig. 15.2). There are about 9,500 living reptile species and 9,900 bird species.

Although most of the species of reptiles that once lived are now extinct, they were highly successful in terms of diversity and duration of existence. Reptiles dominated the earth for millions of years during the Mesozoic era. Turtles do not have any temporal fenestra (temporal openings in the skull), and as such have been historically considered primitive anapsids. The rest of the living Sauropsids have two temporal fenestra, and constitute the subclass Diapsida. Although some recent phylogenetic analyses indicate that modern turtles arose from within the Diapsida, most taxonomists are awaiting more corroborative evidence before they change their taxonomic scheme to reflect evolutionary history. Similarly, despite the fact that birds form a subclade within the infraclass Archosauromorpha, they continue to be ranked by most ornithologists as a separate class (Aves). For convenience we use the most commonly accepted classification scheme, which recognizes two sauropsid subclasses: the Anapsida, which contains extinct 'proto-reptiles' and turtles, and the Diapsida, which contains the majority of 'reptiles,' dinosaurs, and birds. The extant Diapsids are further subdivided into two orders in the infraclass Lepidosauromorpha, Sphenodontia (= Rhynchocephalia; tuataras), and Squamata (lizards and snakes), and one order, Crododilia (alligators, crocodyles, gavials, and caimans) within the infraclass Archosauromorpha (see fig. 15.2). Birds (class Aves) are discussed in further detail in Chapter 16.

Except for a few species of sea snakes and sea turtles, reptiles complete their entire life cycle on land and are considered terrestrial vertebrates. Adaptations that permit utilization of non-aqueous habitats include *keratinized scales* that protect the animal from desiccation, *internal fertilization*, and *amniotic eggs* protected by leathery shells. *Keratin* is a protein within scales that contributes to their strength and waterproofing capabilities. Internal fertilization prevents desiccation of the sperm and requires a male intromittent organ. Male turtles and tortoises, for example, have a *penis,* and male snakes and lizards have two *hemipenes* to transport ejaculated sperm to the female cloaca. The *amnion* is a thin embryonic membrane that insures that development is within the homogeneous and protective environment of *amniotic fluid.* Because the embryonic development of reptiles, birds, and mammals is within amniotic fluid, these groups of vertebrate animals are sometimes referred to as *amniotes.* The anamniotes are the cyclostomes, fishes, and amphibians, or the vertebrates that lack an amnion and amniotic fluid.

Some of the characteristics of reptiles include:

1. A dry scaly skin that is keratinized to protect against desiccation.

2. Internal fertilization usually involving an intromittent, or copulatory, organ to maximize the survival of the gametes.

3. Development of the shelled amniotic egg, in which the embryo forms in a protective fluid environment.

4. The lack of a larval stage of development.

5. Except for snakes and some lizards, the body is supported by girdles and appendages that permit effective terrestrial locomotion. Snakes have secondarily lost their limbs through a selective evolutionary process.

6. *Single occipital condyle*—The articulation between the skull and the atlas vertebra of the vertebral column is through a single bony protuberance. Reptiles and birds have this arrangement, whereas amphibians and mammals have double occipital condyles.

7. *Ectothermy*—Of the vertebrates, only the birds and mammals are homeothermous (metabolically generate a consistent internal body temperature), or endothermous. All other vertebrates are ectothermous and are dependent upon environmental factors to ensure a functional body temperature. Some reptiles, however, are able to maintain a relatively consistent internal body temperature through behavioral adaptations.

8. Enlarged sensory perception portions of the brain and twelve pairs of cranial nerves.

Figure 15.1 A tuatara, *Sphenodon punctatus*. Endemic to New Zealand, tuataras are the only surviving members of order Rhynchocephalia, which flourished around 200 million years ago.

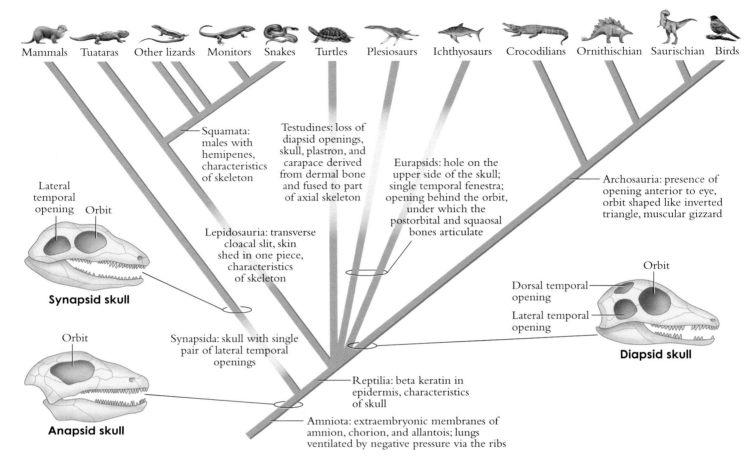

Figure 15.2 Phylogenetic relationships and classification of Reptilia (= Sauropsida).

Class Reptilia (= Sauropsida)

Figure 15.3 Example reptiles include (starting on previous page): (a) a Galapagos green sea turtle, *Chelonia mydas agassisi,* (b) a pair of desert spiny lizards, *Sceloporus magister,* (c) a Jameson's mamba, *Dendroaspis jamesoni,* (d) a green basilisk or plumed basilisk, *Basiliscus plumifrons,* (e) a western coachwhip snake, *Masticophis flagellum,* (f) a spiny soft-shell turtle, *Apalone spinifera,* (g) a gopher tortoise, *Gopherus agassizii,* (h) an American alligator, *Alligator mississippiensis,* (i) a Komodo dragon, *Varanus komodoensis,* (j) a panther chameleon, *Furcifer pardalis,* (k) a Galapagos marine iguana, *Amblyrhynchus cristatus,* and (l) a ring-neck snake, *Diadophis punctatus.*

Figure 15.4 Examples of the four crocodilian types include: (a) American alligator, *Alligator mississippiensis*, (b) Johnston's freshwater crocodile, *Crocodylus johnstoni*, (c) Cuvier's dwarf caiman, *Paleosuchus palpebrosus*, and (d) a gharial, *Gavialis gangeticus*.

Subclass Anapsida - Order Testudines

Figure 15.5 (a) A Hawaiian green turtle, *Chelonia mydas*, and (b) a Galapagos tortoise, *Chelonoidis nigra*, are just two of the many members of order testudines that are threatened or endangered.

Figure 15.6 An Aldabra giant tortoise, *Dipsochelys dussumieri*.

Figure 15.7 A red-eared slider, *Trachemys scripta elegans*.

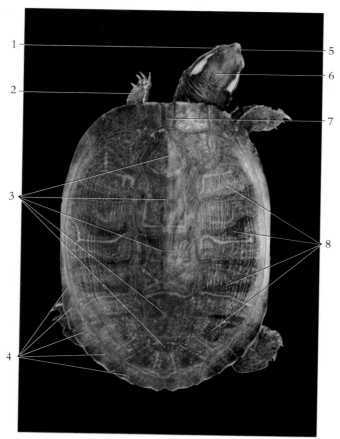

Figure 15.8 A dorsal view of a turtle.
1. Eye
2. Pentadactyl foot
3. Vertebral scales
4. Marginal scales (encircle the carapace)
5. Nostril
6. Head
7. Nuchal scale
8. Costal scales

Figure 15.9 A ventral view of a turtle.
1. Gular scales
2. Humeral scales
3. Pectoral scales
4. Abdominal scales
5. Femoral scales
6. Anal scales
7. Tail

Figure 15.10 The skull of a turtle.

1. Parietal bone
2. Supraoccipital bone
3. Postorbital bone
4. Jugal bone
5. Quadratojugal bone
6. Exoccipital bone
7. Quadrate bone
8. Supraangular bone
9. Articular bone
10. Angular bone
11. Frontal bone
12. Prefrontal bone
13. Palatine bone
14. Premaxilla
15. Maxilla
16. Beak
17. Dentary

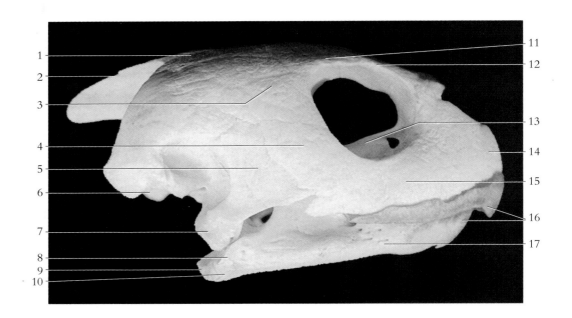

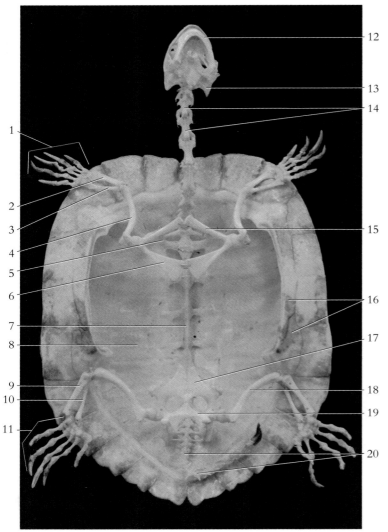

Figure 15.11 The skeleton of a turtle. (The plastron is removed.)

1. Manus (carpal bones, metacarpal bones, phalanges)
2. Radius
3. Ulna
4. Humerus
5. Procoracoid
6. Scapula
7. Vertebra
8. Rib
9. Tibia
10. Fibula
11. Pes (tarsal bones, metatarsal bones, phalanges)
12. Dentary
13. Articular
14. Cervical vertebrae
15. Acromion process of scaphoid
16. Dermal plate of carapace
17. Pubis
18. Femur
19. Ischium
20. Caudal vertebrae

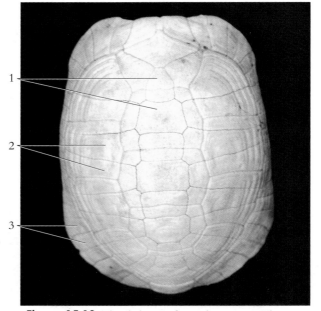

Figure 15.12 The skeleton of a turtle carapace. The epidermally derived keratinous scutes have been removed, exposing the underlying dermal bones. The margins of the scutes and the bones' sutures do not align with one another.

1. Vertebral
2. Costal
3. Marginal

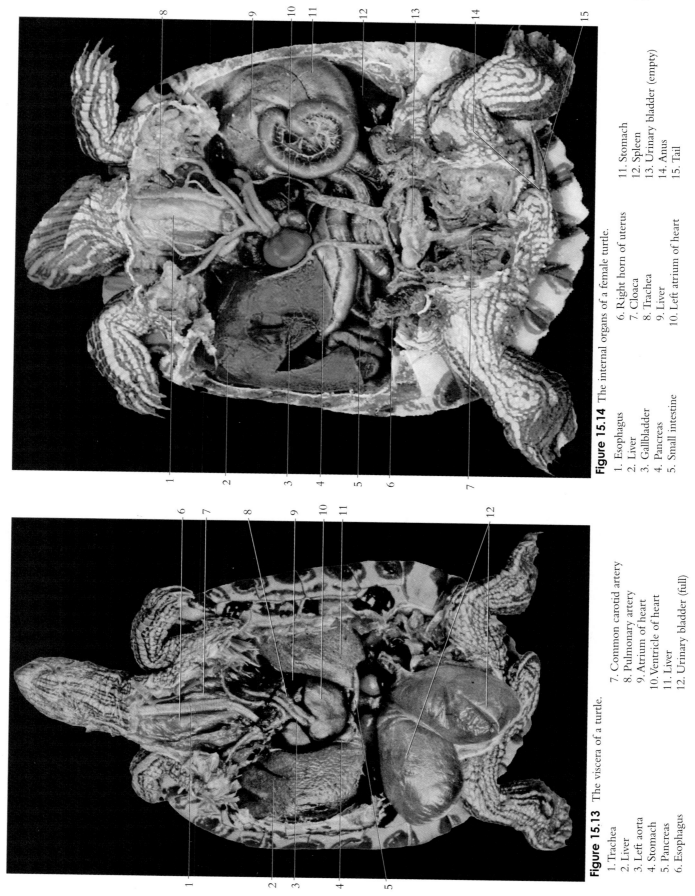

Figure 15.13 The viscera of a turtle.

1. Trachea
2. Liver
3. Left aorta
4. Stomach
5. Pancreas
6. Esophagus
7. Common carotid artery
8. Pulmonary artery
9. Atrium of heart
10. Ventricle of heart
11. Liver
12. Urinary bladder (full)

Figure 15.14 The internal organs of a female turtle.

1. Esophagus
2. Liver
3. Gallbladder
4. Pancreas
5. Small intestine
6. Right horn of uterus
7. Cloaca
8. Trachea
9. Liver
10. Left atrium of heart
11. Stomach
12. Spleen
13. Urinary bladder (empty)
14. Anus
15. Tail

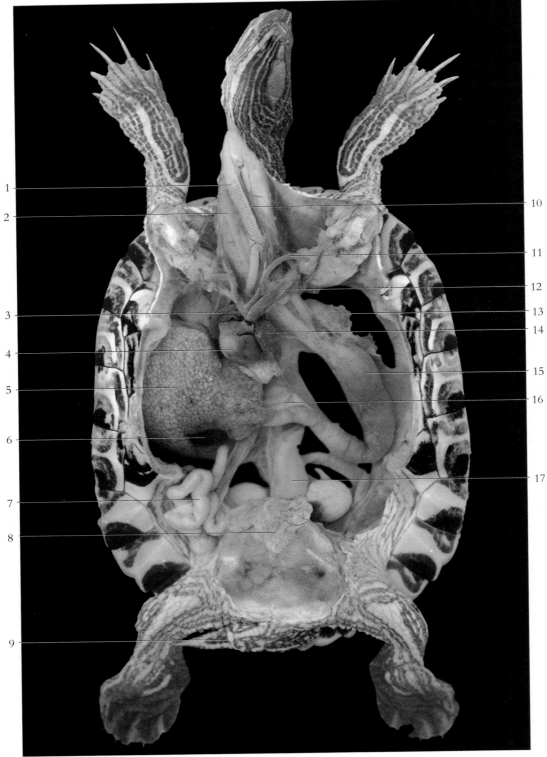

Figure 15.15 A ventral view of the internal anatomy of the turtle.

1. Trachea
2. Esophagus
3. Brachiocephalic trunk
4. Ventricle of heart
5. Liver
6. Gallbladder
7. Small intestine
8. Urinary bladder
9. Anus
10. Common carotid artery
11. Subclavian artery
12. Aortic arch
13. Lung
14. Auricle of heart
15. Stomach
16. Pancreas
17. Colon

Subclass Diapsida - Order Squamata

Figure 15.16 Example squamates include: (a) a Galapagos land iguana, *Conolophus subcristatus*, (b) a mountain kingsnake, *Lampropeltis pyromelana*, (c) a California king snake, *Lampropeltis getula*, (d) an eastern glass lizard, *Ophisaurus ventralis*, a legless lizard, (e) a gila monster, *Heloderma suspectum*, a venomous lizard, and (f) a horned lizard, *Phrynosoma platyrhinos*.

Figure 15.17 Color is used by (a) the Arizona coral snake, *Micruroides euryxanthus*, to warn would-be predators that it is venomous. The scarlet kingsnake, *Lampropeltis triangulum elapsoides,* (b) a nonvenomous snake, mimics the colors of the coral snake to trick predators into leaving it alone. Knowing the pattern, red-yellow-black-yellow versus red-black-yellow-black, can be useful in determining whether a snake is a venomous coral snake or not.

Figure 15.18 (a) A California king snake, *Lampropeltis getula*, in process of ecdysis, or shedding its skin, and (b) a closer view.

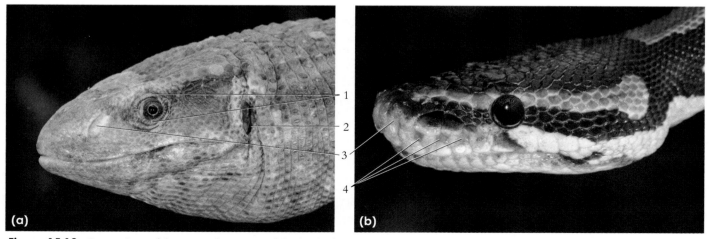

Figure 15.19 Comparison of the external anatomy of the head of (a) the savannah monitor lizard (*Varanus exanthematicus*) and (b) the ball python (*Python regius*). Lizards have eyelids and external ears, where as these structures are lacking in snakes. (Notice the heat pit receptors, which are characteristic of pythons and pit vipers and are adaptive for predation on warm-blooded vertebrates.)

1. Eyelids   2. External ear   3. Nostril   4. Heat pits

Figure 15.20 Hatching California king snakes, *Lampropeltis getula*. Most snakes are oviparous, meaning that they lay eggs such as these. Some snakes, including American pit vipers, are ovoviviparous, giving birth to well-developed young.

Figure 15.21 A garter snake, *Thamnophis sirtalis*, extending its tongue. Reptiles use their tongue in conjunction with the Jacobson's organ or vomeronasal organ, an auxiliary olfactory organ, to aid with smell.

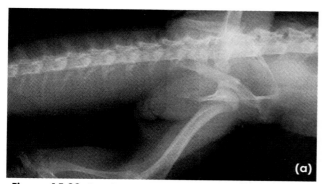
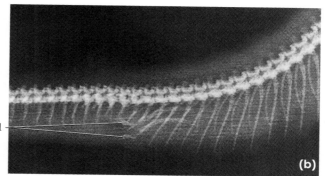

Figure 15.22 A radiograph of the pelvic region of a savannah monitor (a) showing a highly developed limb. Compare this to the radiograph of the pelvic region of a boa (b) showing the vestigial pelvic girdle.
1. Vestigial pelvic girdle

Figure 15.23 A radiograph of the skeleton of the argus monitor, *Varanus panoptes*.

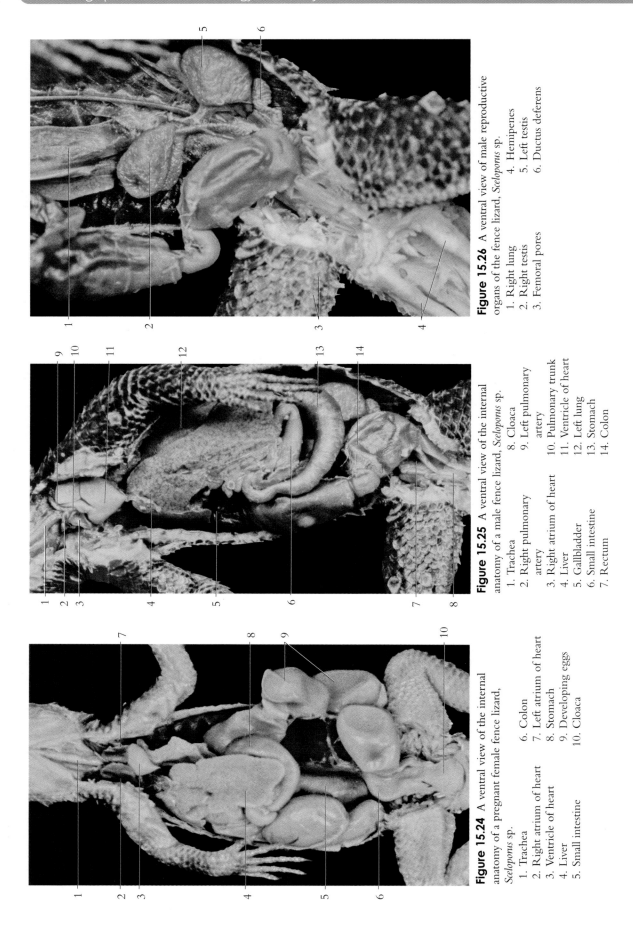

Figure 15.24 A ventral view of the internal anatomy of a pregnant female fence lizard, *Sceloporus* sp.

1. Trachea
2. Right atrium of heart
3. Ventricle of heart
4. Liver
5. Small intestine
6. Colon
7. Left atrium of heart
8. Stomach
9. Developing eggs
10. Cloaca

Figure 15.25 A ventral view of the internal anatomy of a male fence lizard, *Sceloporus* sp.

1. Trachea
2. Right pulmonary artery
3. Right atrium of heart
4. Liver
5. Gallbladder
6. Small intestine
7. Rectum
8. Cloaca
9. Left pulmonary artery
10. Pulmonary trunk
11. Ventricle of heart
12. Left lung
13. Stomach
14. Colon

Figure 15.26 A ventral view of male reproductive organs of the fence lizard, *Sceloporus* sp.

1. Right lung
2. Right testis
3. Femoral pores
4. Hemipenes
5. Left testis
6. Ductus deferens

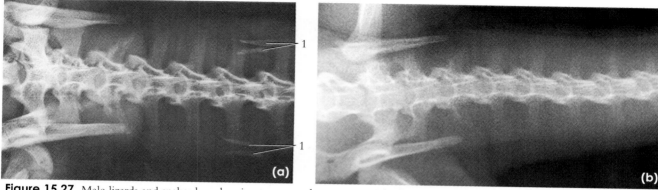

Figure 15.27 Male lizards and snakes have hemipenes as copulatory organs. The hemipene seen in a radiograph of a male (a) crocodile monitor, *Varanus salvadorii*. As seen in a radiograph, a female (b) lacks a hemipenis. The female cloaca is the receptacle of the everted male hemipenis during copulation.
1. Sheaths of hemipenes

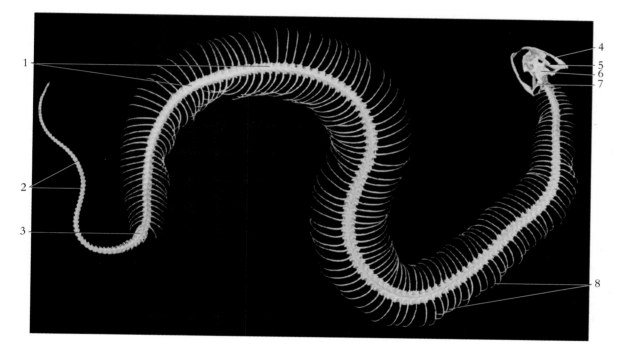

Figure 15.28 The skeleton of a snake (python).
1. Trunk vertebrae
2. Caudal vertebrae
3. Vestigial pelvic girdle
4. Dentary
5. Supratemporal bone
6. Parietal
7. Quadrate bone
8. Ribs

Figure 15.29 The oral region of a spitting cobra, *Naja* sp., showing its venom gland, which is a modified salivary gland; and its fang, which is a modified maxillary tooth.
1. Eye
2. Venom gland
3. Fang sheath
4. Fang

Figure 15.30 The overlapping arrangement of snake scales.

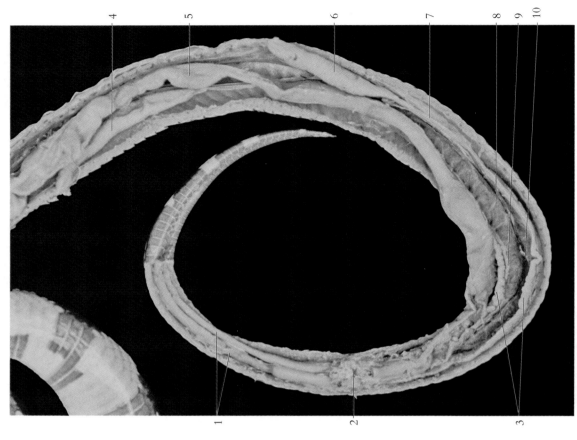

Figure 15.31 A ventral view of the internal anatomy of a female water moccasin, *Agkistrodon piscivorus*.

1. Jugular vein
2. Trachea
3. Common carotid artery
4. Esophagus
5. Aortic arch
6. Auricle of heart
7. Ventricle of heart
8. Lung
9. Hepatic portal vein
10. Liver
11. Dorsal aorta
12. Stomach
13. Anus
14. Colon
15. Kidney
16. Oviduct
17. Eggs
18. Small intestine
19. Pancreas
20. Duodenum
21. Abdominal vein

Figure 15.32 A ventral view of the internal anatomy of a male California king snake, *Lampropeltis getula*. Note that the testes and kidneys are staggered.

1. Hemipenes
2. Anus
3. Ductus deferens
4. Right testis
5. Intestine
6. Left testis
7. Ductus deferens
8. Right kidney
9. Ureter
10. Left kidney

Aves

Chapter 16

Birds are homeothermous (warm-blooded) chordate animals within the class Aves. There are about 9,990 species of birds. Class Aves is divided into two subclasses: *Archaeornithes* and *Neornithes*. Subclass *Archaeornithes* includes the fossil, *Archaeopteryx*, and subclass *Neornithes* includes all other birds, fossil and modern.

Birds are closely related to reptiles and seem to have evolved from a basic archosaurian stock of quadrupedal animals in the Jurassic period some 140 million years ago. Crocodiles and alligators are also archosaurs and are the closest living relatives of birds. Crocodiles and alligators, however, have specializations for an amphibious life, whereas birds have specializations for flight and bipedalism (support on two appendages).

Some of the characteristics of birds include:

1. *Feathers*—Derived from the epidermis of the skin, feathers provide protection, insulation necessary for homeothermia, and lightweight body surface for flight.

2. *Beaks*—Derived from the epidermis of the skin, beaks are highly adapted for a variety of functions, including obtaining food, building nests, preening, defense, and mating.

3. *Thin skin*—The skin of birds is thin, vascular, and has few glands. Oil-secreting glands line the outer ear canal and encircle the vent of the cloaca in some birds. The uropygial gland at the base of the tail is also an oil-secreting gland that is used in preening and is especially well-developed in aquatic birds.

4. *Hollow bones*—The sternum, vertebrae, and several of the long bones of birds are pneumatic, meaning that they have a hollow, air-filled core. Pneumatic bones contribute to the skeletal system of a bird being lightweight and yet strong and durable.

5. *Fusion of bones*—A reduction in the number of bones also contributes to the lightweight body of a bird, while still providing structural support. Other bones have specializations to accommodate flight. The bones of the forelimbs, for example, are adapted for flight, and the ribs have uncinate processes that strengthen the rib cage and provide an increased surface area for muscle attachment.

6. *Sclerotic bones*—A ring of bones within the anterior portion of the eyeball of birds and certain reptiles provides structural support.

7. *Single occipital condyle*—The articulation between the skull and the atlas vertebra of the vertebral column is through a single occipital condyle rather than a double occipital condyle, as is the arrangement in mammals.

8. *Skeletal muscles positioned primarily upon the torso*—The large and powerful flight muscles are positioned upon the sternum and rib cage along the ventral side of the thorax. In this position, these heavy muscles provide ballast to the bird during flight. The muscles that move the distal portions of the appendages are positioned proximally and act upon their insertion attachment points by long, lightweight tendons.

9. *Four-chambered heart with a right aortic arch*—Although mammals also have a four-chambered heart, their aortic arch extends to the left rather than to the right. A four-chambered heart and separate pulmonary (to the lungs) and systemic (to the body) blood flow are important adaptations for homeothermia.

10. *Single functional ovary*—Although present during embryonic development, in most birds the right ovary and oviduct become vestigial so that only the left genital structures are functional.

11. *Oviparous*—Birds have internal fertilization and are egg-laying, or oviparous.

12. *Air sacs*—A series of air sacs (generally nine) is situated in various parts of the body of a bird and functions principally as air reservoirs. The air sacs are connected to the bronchi of the lungs.

13. *Well-developed brain*—Although non-convoluted, the large cerebrum of the avian brain enables an array of intricate instinctive behaviors, including courtship, nest building, and migration.

Figure 16.1 A hooded merganser, *Lophodytes cucullatus*, is the smallest of the North American mergansers.

Class Aves

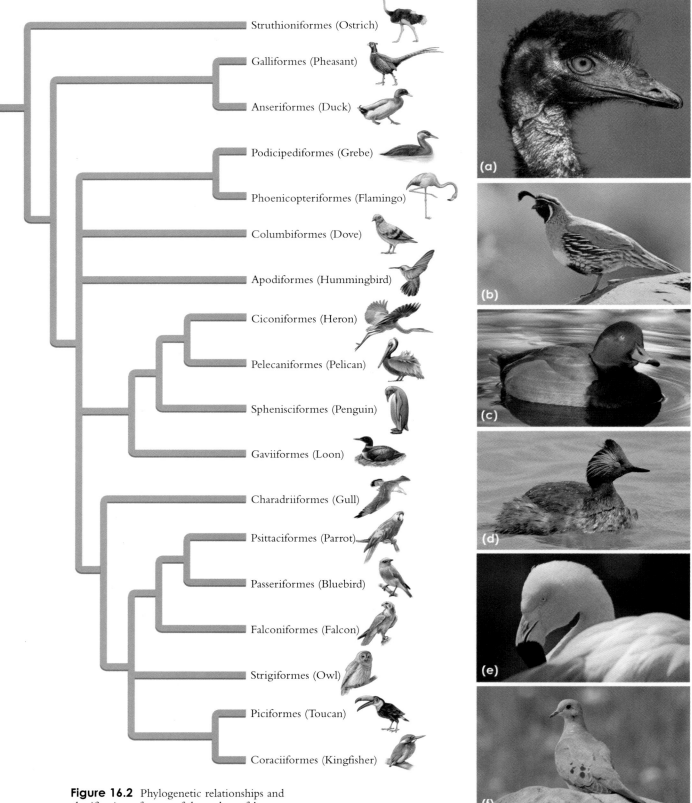

Figure 16.2 Phylogenetic relationships and classification of some of the orders of Aves.

Figure 16.3 Examples of Aves and their associated order include (starting on previous page): (a) an emu, *Dromaius novaehollandiae*, order Struthioniformes (b) a California quail, *Callipepla californica*, order Galliformes, (c) a redhead duck, *Aythya americana*, order Anseriformes, (d) an eared grebe, *Podiceps nigricollis*, order Podicipediformes, (e) a Chilean flamingo, *Phoenicopterus chilensis*, order Phoenicopteriformes, (f) a mourning dove, *Zenaida macroura*, order Columbiformes, (g) a sparkling violetear hummingbird, *Colibri coruscans*, order Apodiformes, (h) a lava heron, *Butorides sundevalli*, order Ciconiformes, (i) a brown pelican, *Pelecanus occidentalis*, order Pelecaniformes, (j) a Galapagos penguin, *Spheniscus mendiculus*, order Sphenisciformes, (k) a pacific loon, *Gavia pacifica*, order Gaviiformes, (l) a Franklin's gull, *Leucophaeus pipixcan*, order Charadriiformes, (m) a blue and gold macaw, *Ara ararauna*, order Psittaciformes, (n) a Brewer's blackbird, *Euphagus cyanocephalus*. order Passeriformes, (o) a peregrine falcon, *Falco peregrinus*, order Falconiformes, (p) a barn owl, *Tyto alba*, order Strigiformes (q) a red-shafted flicker or northern flicker, *Colaptes auratus cafer*, order Piciformes, and (r) a belted kingfisher, *Megaceryle alcyon*, order Coraciformes.

Figure 16.4 Example beaks and their associated uses include: (a) nectar feeding, a broad-tailed hummingbird, *Selasphorus platycercus*, (b) grain eating, a lazuli bunting, *Passerina amoena*, (c) dip netting, an American white pelican, *Pelecanus erythrorhynchos*, (d) flesh eating, an American bald eagle, *Haliaeetus leucocephalus*, (e) insect eating, a western tanager, *Piranga ludoviciana*, (f) scavenging, a king vulture, *Sarcoramphus papa*, (g) filter feeding, a Chilean flamingo, *Phoenicopterus chilensis*, (h) generalist, a ring-billed gull, *Larus delawarensis*, (i) chiseling, red-shafted flicker or northern flicker, *Colaptes auratus cafer,* (j) probing, an American avocet, *Recurvirostra americana*, (k) spearing, an American darter or anhinga, *Anhinga anhinga*, and (l) nut cracking, a scarlet macaw, *Ara macao*.

Figure 16.5 Within the animal kingdom birds are some of the best examples of sexual dimorphism, the morphological difference between males and females of the same species. (a) A pair of wood ducks, *Aix sponsa*, passes a crabapple during courtship. (b) a male Indian peafowl, *Pavo cristatus*, known commonly as a peacock, and (c) a female known as a peahen.

Figure 16.6 (a) A red-tailed hawk, *Buteo jamaicensis*, has long broad wings used for soaring on thermals. Its wing beat is somewhat slow with periods of gliding inbetween several wing beats, while (b) the American kestrel, *Falco sparverius*, a falcon, has narrow, long and pointed wings best suited for direct fast flight. Its wing beats are fast with relatively no breaks for gliding.

Figure 16.7 Plumage plays many roles. For (a) a great horned owl, *Bubo virginianus*, it aids in camouflage, while (b) the Bullock's oriole, *Iterus bullockii*, uses it for territorial display and attracting attention.

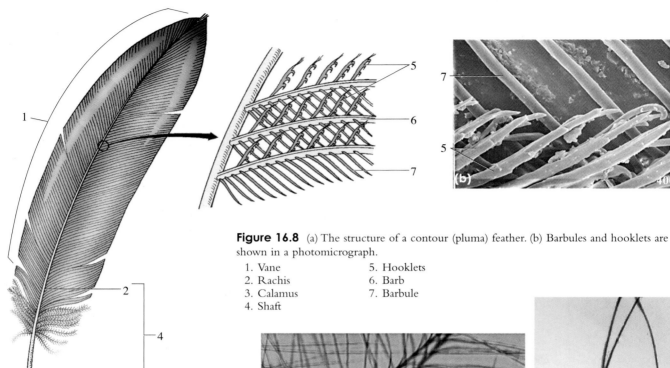

Figure 16.8 (a) The structure of a contour (pluma) feather. (b) Barbules and hooklets are shown in a photomicrograph.

1. Vane
2. Rachis
3. Calamus
4. Shaft
5. Hooklets
6. Barb
7. Barbule

Figure 16.9 Pluma (*pl. plumae*) are contour feathers that include the quill feathers and flight feathers. Quill feathers cover most of the body. Flight feathers are confined to the wings as remiges and to the tail as retrices (scale in mm).

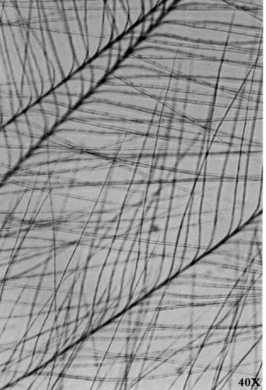

Figure 16.10 The plumula (*pl. plumulae*) is a soft insulating feather that lacks a rachis and hooklets. Also called down feathers, plumulae are abundant on the breast and abdomen of a bird. Plumulae are shown in this micrograph.

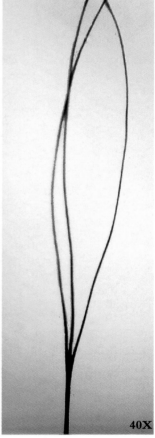

Figure 16.11 The filoplume (*pl. filoplumes*) is a hairlike feather consisting of a single rachis and either no barbs or just a few at its distal tip. Filoplumes are scattered throughout the plumae. A filoplume is shown in this micrograph.

Figure 16.12 A rock dove, *Columba* sp.

Figure 16.13 A broad-tailed hummingbird, *Selasphorus platycercus*.

Figure 16.16 A great blue heron, *Ardea herodias*.

Figure 16.14 A wood duck, *Aix sponsa*.

Figure 16.15 A nighthawk, *Chordeiles minor*.

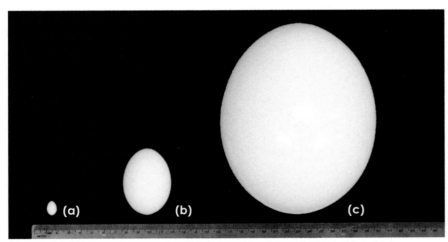

Figure 16.17 Size comparison of bird eggs; (a) a hummingbird, (b) a chicken, and (c) an ostrich (scale in mm).

Figure 16.18 A gila woodpecker, *Melanerpes uropygialis*, "anting." The woodpecker allows ants to crawl through its feathers and eat any parasites that the bird might have.

Figure 16.19 Flocking to a roosting site is an important social behavior in many species of birds.

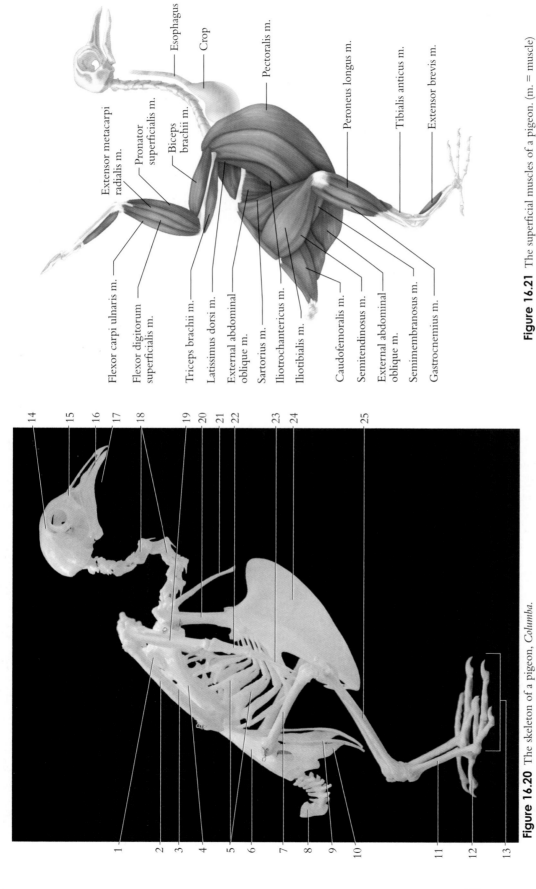

Figure 16.20 The skeleton of a pigeon, *Columba*.

1. Scapula
2. Ulna
3. Radius (behind ulna)
4. Humerus
5. Ribs
6. Ilium
7. Femur
8. Pygostyle
9. Pubis
10. Ischium
11. Tarsometatarsal bone
12. Digit 1
13. Phalanges
14. Cranium
15. Sclerotic bone
16. Premaxilla
17. Dentary
18. Cervical vertebrae
19. Carpometacarpal bones
20. Coracoid bone
21. Furcula
22. Phalanges
23. Phalanx of 3rd digit
24. Keel of sternum
25. Tibiotarsal bone

Figure 16.21 The superficial muscles of a pigeon. (m. = muscle)

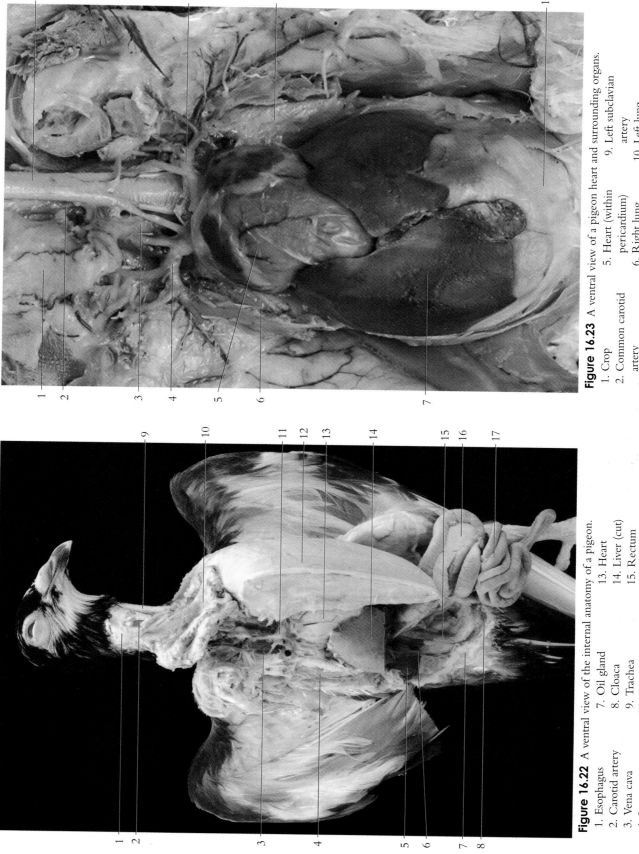

Figure 16.23 A ventral view of a pigeon heart and surrounding organs.

1. Crop
2. Common carotid artery
3. Esophagus
4. Right subclavian artery
5. Heart (within pericardium)
6. Right lung
7. Liver
8. Trachea
9. Left subclavian artery
10. Left lung
11. Greater omentum

Figure 16.22 A ventral view of the internal anatomy of a pigeon.

1. Esophagus
2. Carotid artery
3. Vena cava
4. Lung
5. Kidney
6. Oviduct
7. Oil gland
8. Cloaca
9. Trachea
10. Crop
11. Aortic arch
12. Pectoralis muscle
13. Heart
14. Liver (cut)
15. Rectum
16. Pancreas
17. Ileum

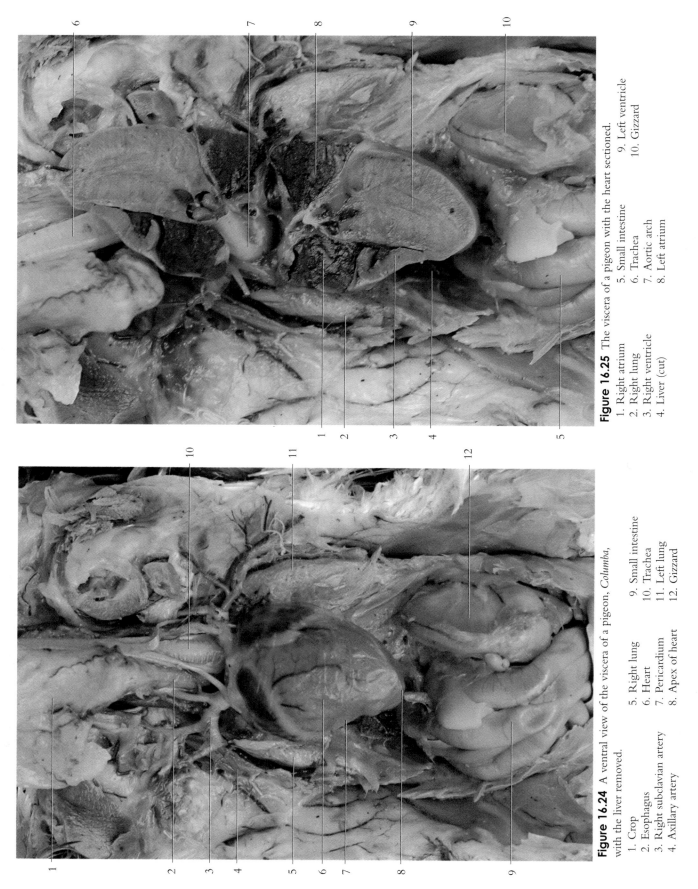

Figure 16.24 A ventral view of the viscera of a pigeon, *Columba*, with the liver removed.

1. Crop
2. Esophagus
3. Right subclavian artery
4. Axillary artery
5. Right lung
6. Heart
7. Pericardium
8. Apex of heart
9. Small intestine
10. Trachea
11. Left lung
12. Gizzard

Figure 16.25 The viscera of a pigeon with the heart sectioned.

1. Right atrium
2. Right lung
3. Right ventricle
4. Liver (cut)
5. Small intestine
6. Trachea
7. Aortic arch
8. Left atrium
9. Left ventricle
10. Gizzard

Mammalia

Chapter 17

Mammals are homeothermous (warm-blooded) chordate, synapsid animals within the class Mammalia. There are about 5,750 species of mammals. Phylogenetic analyses of mammalian lineages is an active discipline and necessitates changes in classification, primarily at higher levels (subclass, infraclass, and order). The current best estimate of phylogenetic relationships supports three independent lineages within the Mammalia: Monotremata (egg-laying mammals), Marsupialia (marsupials), and Eutheria (placental mammals).

Monotremes have a cloaca, or single duct that drains the urinary, defecatory, and reproductive systems. They lay eggs, but they also lactate and excrete milk from mammary glands under the skin (they do not have nipples). Marsupials give birth to poorly developed young and have a front pouch with nipples where young nurse and develop, but marsupials do not have a placenta, which evolved in the eutherians and serves to protect the developing embryo from the maternal immune system. Eutherians also have a corpus callosum, which connects the right and left hemispheres of the brain. The earliest mammals arose in the middle of the late Jurassic period and underwent rapid diversification following the demise of the dinosaurs.

Some of the characteristics of mammals include:

1. *Hair*—Derived from the epidermis of the skin, hair provides protection and insulation necessary for homeothermia. Some mammals are sparsely haired, including marine mammals and humans.

2. *Mammary glands*—Modified sweat glands, the mammary glands produce milk upon parturition (birth) for suckling the young.

3. *Four-chambered heart with a left aortic arch*—Although birds also have a four-chambered heart, their aortic arch extends to the right rather than to the left. A four-chambered heart and separate pulmonary (to the lungs) and systemic (to the body) blood flow support a high metabolism and thus are important adaptations for homeothermia.

4. *Well-developed brain with a convoluted cerebrum*—Not only is the brain of mammals relatively larger than that of other vertebrates, but the cerebrum is convoluted. Convolutions (gyri and sulci) increase the surface area where cell bodies of neurons are located.

5. *Three auditory ossicles*—The auditory ossicles (bones) function as levers in amplifying sound waves passing through the middle-ear chamber.

6. *Double occipital condyle*—Also found in amphibians, the double occipital condyle provides support of the head on the vertebral column and permits extensive flexibility.

7. *Single dentary bone*—The lower jaw, or mandible, consists of one bone that articulates at the skull at the temporomandibular joint.

8. *Seven cervical vertebrae*—Although a few species of mammals do not have seven cervical (neck) vertebrae, this is generally considered a diagnostic mammalian trait.

9. *Heterodont dentition*—Teeth differ in structure and are adapted to handle food in different ways. The distinct teeth include incisors, canines, premolars, and molars.

10. *Pinnea* (fleshy outer ears) and *movable eyelids*—Each pinna funnels sound waves to the auditory tube; the movable eyelids protect the anterior surface of the eyeball and keep it from drying out.

11. *Muscular diaphragm*—Separating the thoracic and abdominal cavities, the muscular diaphragm is the principal muscle of inspiration (inhalation). An effective respiratory system is essential for maintaining homeothermia.

12. *Nonnucleated, biconcave red blood cells*—The red blood cells, or erythrocytes, are produced in red bone marrow and transport oxygen to each of the living cells within the mammalian body.

13. *Placental attachment of fetus*—Mammals have internal fertilization; prenatal development is in a uterus with placental attachment (except in monotremes), and fetal membranes (amnion, chorion, allantois) protect and sustain the fetus.

The highly specialized mammalian brain has enabled these animals to learn and adapt in response to experiences and to evolve intricate social structures within species and populations.

Figure 17.1 The bottlenose dolphin, *Tursiops truncatus*, is a highly social marine mammal.

Class Mammalia

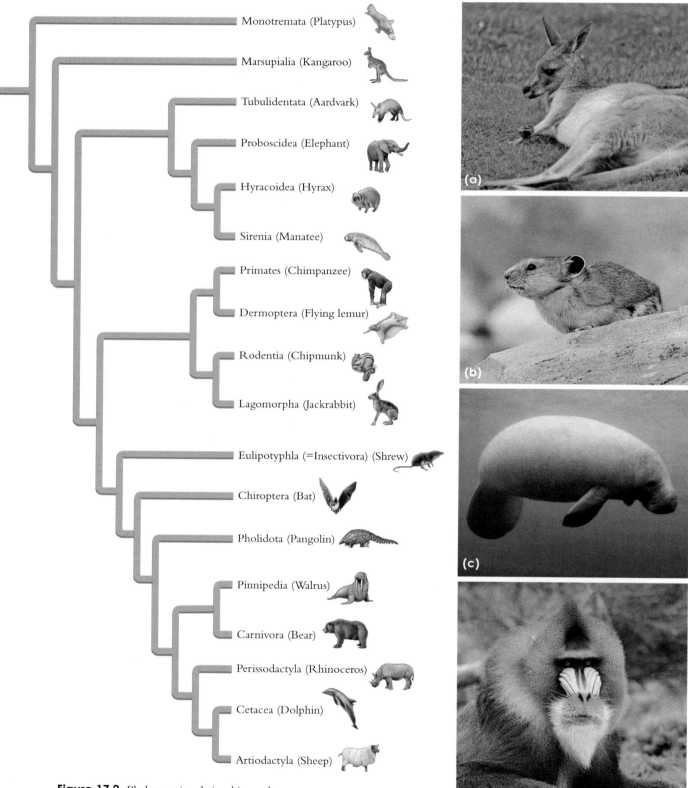

Figure 17.2 Phylogenetic relationships and classification of some of the orders of Mammalia.

Figure 17.3 Examples of Mammalia and their associated order include (starting on previous page): (a) an eastern grey kangaroo, *Macropus giganteus*, order Marsupialia (b) a Hyrax, *Ochotona princeps,* order Hyracoidea, (c) a West Indian manatee, *Trichechus manatus*, order Sirenia, (d) a mandrill, *Mandrillus sphinx,* order Primates, (e) a Utah prairie dog, *Cynomys parvidens*, order Rodentia, (f) a cottontail rabbit, *Sylvilagus audubonii,* order Lagomorpha, (g) Malaysian fruit bats, *Pteropus hypomelanus*, order Chiroptera, (h) a common seal or harbor seal, *Phoca vitulina*, suborder Pinnipedia, (i) a grizzly bear, *Ursus arctos horribilis,* order Carnivora, (j) a black rhinoceros or hook-lipped rhinoceros, *Diceros bicornis*, order Perissodactyla, (k) a bottlenose dolphin, *Tursiops truncatus*, order Cetacea, and (l) a mule deer, *Odocoileus hemionus*, order Artiodactyla.

Figure 17.4 The capybara, *Hydrochoerus hydrochaeris*, is the largest living rodent. It can grow to over 4 feet in length and weigh over 175 pounds.

Figure 17.5 The lesser long-nosed bat, *Leptonycteris yerbabuenae*. Bats are the only true flying mammal. Some bat species are critical for pollination of certain plant species.

Figure 17.6 The pronghorn antelope, *Antilocapra americana*, is cited as the second fastest land animal behind the cheetah. It is reported that it can reach speeds up to 70 miles per hour.

Figure 17.7 The ringtailed lemur, *Lemur catta*, is a primate. There are approximately 100 species of lemurs, all of which are restricted to the island of Madagascar.

Figure 17.8 The American bison, *Bison bison*, is the largest living land animal in North America.

Figure 17.9 The western lowland gorilla, *Gorilla gorilla gorilla,* is the smallest subspecies of gorilla. An adult male can reach 6 feet tall and weigh as much as 600 pounds. They are critically endangered.

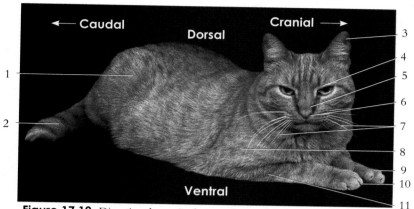

Figure 17.10 Directional terminology and superficial structures in a cat (quadrupedal vertebrate).
1. Thigh
2. Tail
3. Auricle (pinna)
4. Superior palpebra (superior eyelid)
5. Bridge of nose
6. Naris (nostril)
7. Vibrissae
8. Brachium
9. Manus (front foot)
10. Claw
11. Antebrachium

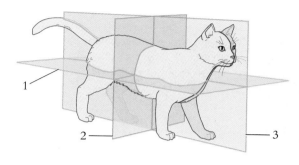

Figure 17.11 The planes of reference in a cat.
1. Coronal plane (frontal plane)
2. Transverse plane (cross-sectional plane)
3. Midsagittal plane (median plane)

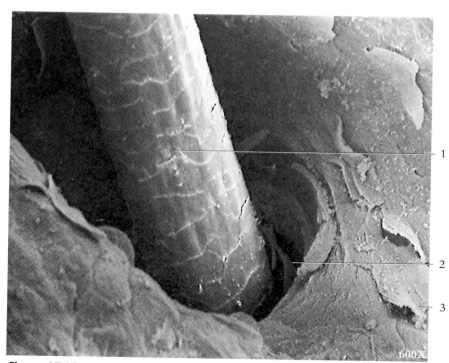

Figure 17.12 A scanning electron micrograph of a hair emerging from a hair follicle.
1. Shaft of hair (note the scale–like pattern)
2. Hair follicle
3. Epithelial cell from stratum corneum

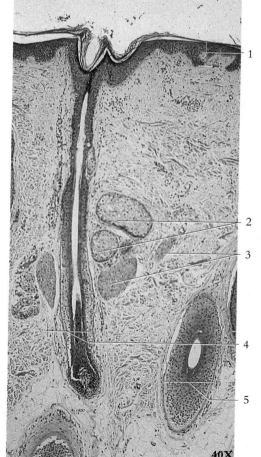

Figure 17.13 A hair follicle.
1. Epidermis
2. Sebaceous glands
3. Arrector pili muscle
4. Hair follicle
5. Hair follicle (oblique cut)

Rat Dissection

The laboratory white rat is a captive-raised rodent that is commercially available for biological and medical experiments and research. White rats are also embalmed and available as dissection specimens in biology, vertebrate biology, and general zoology laboratories. Before the muscles and viscera of a rat can be studied, the specimen's skin has to be removed according to the following suggested guidelines.

1. Place the rat on a dissecting tray dorsal side up. Using a sharp scalpel, make a short, shallow incision through the skin across the neck. With your scissors, continue a dorsal midline incision to about two inches onto the tail. Sever the tail with bone shears and discard.
2. Make a shallow incision around the neck and down each front leg to the paws. Continue a circular incision around each wrist. Beginning at the base of the tail, make incisions down each of the hind legs to the ankles. Make a circular cut around each ankle.
3. Carefully remove the skin, using your fingers or a blunt probe to separate the skin from underlying connective tissue. Where it is necessary to use a scalpel, keep the cutting edge directed toward the skin away from the muscle. If your specimen is a male, make an incision around the genitalia, leaving the skin intact. If your specimen is a female, the mammary glands will appear as longitudinal, glandular masses along the ventral side of the abdomen and thorax. They should be removed with the skin.
4. After the specimen is skinned, remove the excess fat and connective tissue to expose the underlying muscles. Make certain that the muscles are separated along their natural boundaries. If a transection of a muscle is necessary, isolate the muscle from its attached connective tissue, and make a clean cut across the belly of the muscle, leaving the origin and insertion intact.
5. At the end of the laboratory period, wrap your specimen in a cloth and store it in a tight, heavy-duty plastic bag. Wet your specimen from time to time with a preservative solution. Caution has to be taken when using a phenol wetting solution, as it is caustic and poisonous if misused or used in a concentrated form.

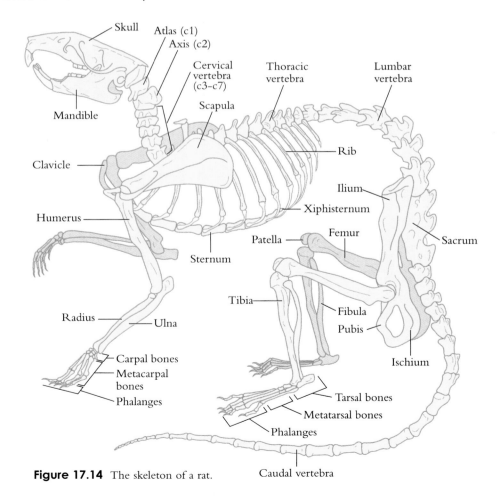

Figure 17.14 The skeleton of a rat.

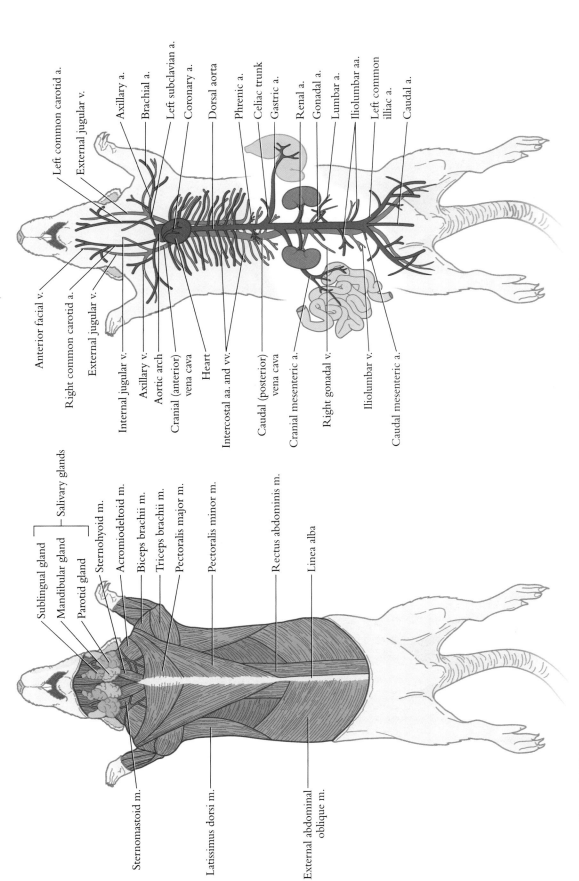

Figure 17.16 The circulatory system of a rat. The arteries are colored red (a. = artery, aa. = arteries; v. = vein, vv. = veins).

Figure 17.15 A ventral view of the superficial musculature of the rat (m. = muscle).

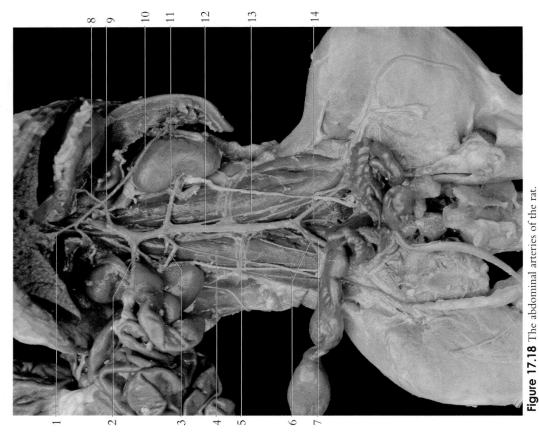

Figure 17.18 The abdominal arteries of the rat.

1. Hepatic artery
2. Right renal artery
3. Cranial mesenteric artery
4. Right testicular artery
5. Right iliolumbar artery
6. Caudal mesenteric artery (cut)
7. Right common iliac artery
8. Gastric artery
9. Celiac trunk
10. Splenic artery
11. Left renal artery
12. Abdominal aorta
13. Left testicular artery
14. Middle sacral artery

Figure 17.17 A ventral view of the rat viscera.

1. Larynx
2. Trachea
3. Right lung
4. Jejunum
5. Right ovary
6. Caecum
7. Esophagus
8. Heart
9. Diaphragm
10. Liver
11. Spleen
12. Ileum

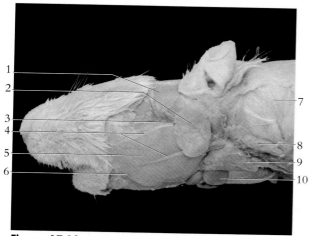

Figure 17.19 The head and neck region of the rat.
1. Temporalis m.
2. Extraorbital lacrimal gland
3. Extraorbital lacrimal duct
4. Facial nerve
5. Masseter m.
6. Parotid duct
7. Cervical trapezius m.
8. Parotid gland
9. Mandibular gland
10. Mandibular lymph node

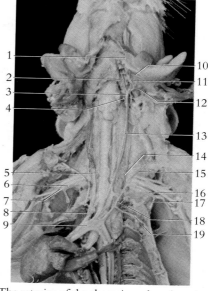

Figure 17.20 The arteries of the thoracic and neck regions of the rat.
1. Facial artery
2. Lingual artery
3. External carotid artery
4. Cranial thyroid artery
5. Right common carotid artery
6. Right axillary artery
7. Right brachial artery
8. Brachiocephalic artery
9. Aortic arch
10. External maxillary artery
11. Internal carotid artery
12. Occipital artery
13. Left common carotid artery
14. Vertebral artery
15. Cervical trunk
16. Lateral thoracic artery
17. Left axillary artery
18. Left subclavian artery
19. Internal thoracic artery

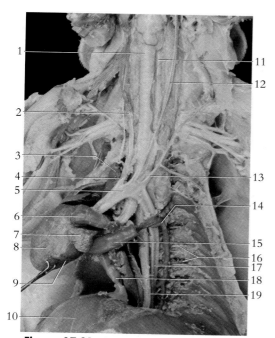

Figure 17.21 The rat heart (reflected) showing the major veins and arteries.
1. Trachea
2. Right common carotid artery
3. Right cranial vena cava (cut)
4. Brachiocephalic trunk
5. Aortic arch
6. Pulmonary trunk
7. Left auricle
8. Left ventricle
9. Coronary vein
10. Diaphragm
11. Esophagus
12. Left common carotid artery
13. Left subclavian artery
14. Azygos vein
15. Coronary sinus (cut)
16. Intercostal artery and vein
17. Aorta
18. Caudal vena cava
19. Esophagus

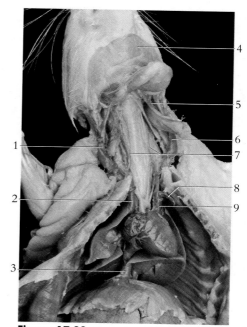

Figure 17.22 The veins of the thoracic and neck regions of the rat.
1. Cephalic vein
2. Right cranial vena cava
3. Caudal vena cava
4. Linguofacial vein
5. Maxillary vein
6. External jugular vein
7. Internal jugular vein
8. Lateral thoracic vein
9. Left cranial vena cava

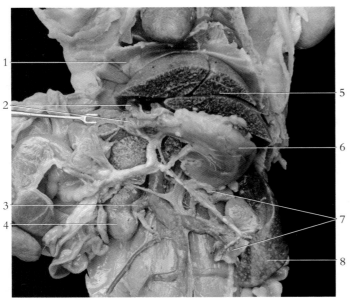

Figure 17.23 The abdominal viscera and vessels of the rat.
1. Diaphragm
2. Biliary and duodenal parts of pancreas
3. Right renal vein
4. Right kidney
5. Liver (cut)
6. Stomach
7. Gastrosplenic part of pancreas
8. Spleen

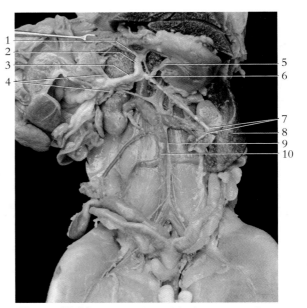

Figure 17.24 The branches of the hepatic portal system.
1. Cranial pancreaticoduodenal vein
2. Hepatic portal vein
3. Cranial mesenteric vein
4. Intestinal branches
5. Gastric vein
6. Gastrosplenic vein
7. Splenic branches
8. Spleen
9. Abdominal aorta
10. Caudal vena cava

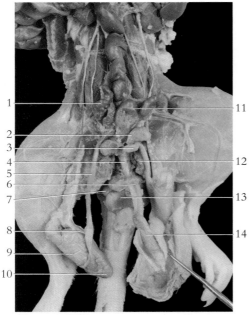

Figure 17.25 The urogenital system of the male rat.
1. Vesicular gland
2. Prostate (ventral part)
3. Prostate (dorsolateral part)
4. Urethra in the pelvic canal
5. Vas (ductus) deferens
6. Crus of penis
7. Bulbourethral gland
8. Head of epididymis
9. Testis
10. Tail of epididymis
11. Urinary bladder
12. Symphysis pubis (cut exposing pelvic canal)
13. Bulbocavernosus muscle
14. Penis

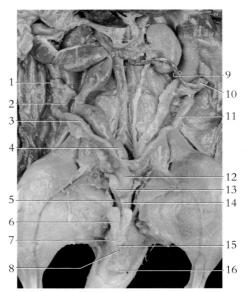

Figure 17.26 The urogenital system of the female rat.
1. Ovary
2. Uterine artery and vein
3. Uterine horn
4. Colon
5. Vagina
6. Vestibular gland
7. Clitoris
8. Vaginal opening
9. Uterine artery and vein
10. Ovary
11. Uterine horn
12. Uterine body
13. Urinary bladder
14. Urethra
15. Urethral opening
16. Anus

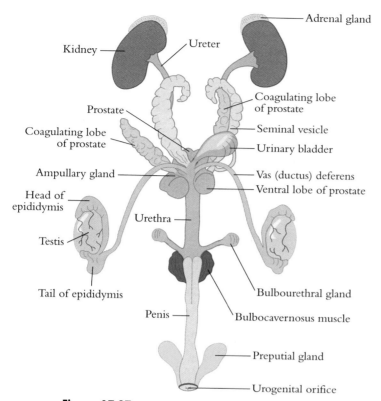

Figure 17.27 The urogenital organs of a male rat.

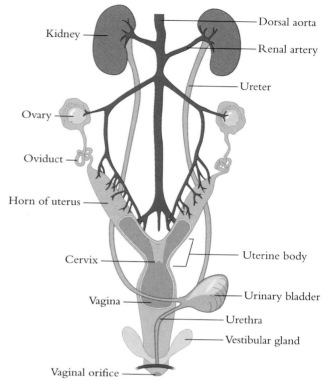

Figure 17.28 The urogenital organs of a female rat.

Fetal Pig Dissection

Fetal pigs are purchased from biological supply houses and are specially prepared for dissection. Excess embalming fluid should be drained from the packaged specimen prior to dissection.

Examine your specimen, and identify the **umbilical cord**. Locate the two rows of **teats** that extend the length of the abdomen. Determine the sex of your specimen. A male has a **scrotal sac** in the pelvic region of the body between the hind legs and a **urogenital opening** just caudal to the umbilical cord. The **penis** can be palpated as a muscular tubular structure just underneath the skin along the midline, proceeding caudally from the urogenital opening. A female has a small, fleshy **genital papilla** projecting from the urogenital opening, which is located immediately ventral to the **anal opening**.

Before the muscles and viscera of a fetal pig can be studied, the specimen's skin has to be removed according to the following suggested guidelines.

1. Place your specimen on a dissecting tray ventral side up. Using a sharp scalpel, make a shallow incision through the skin, extending from the chin caudally to the umbilical cord. Carefully continue your cut around one side of the umbilical cord. If your specimen is a male, make a diagonal cut from the umbilical cord to the scrotum. If a female, continue a midventral incision from the umbilical cord to the genital papilla. Make an incision around the genitalia and tail.
2. From the midventral incision, extend an incision down the medial surfaces of the front legs to the hooves, and then do the same for the skin of the hind legs. Make circular incisions around each of the hooves. Following the ventral borders of the lower jaws, make extended cuts from the chin dorsolaterally to just below the ears.
3. Grasp the cut edge of the skin and carefully remove it from your specimen. If the skin is difficult to remove, grasp the cut edge of the skin with one hand and push on the muscle with the other hand.
4. After the specimen is skinned, the muscles can be seen more easily if the moisture on them is sponged away with a paper towel. The muscles of a fetal pig are extremely delicate, and as you proceed to dissect your specimen, make certain that you separate the muscles along their natural boundaries. When transection of a muscle is necessary, carefully isolate the muscle from its attached connective tissue, and make a clean cut across the belly of the muscle, leaving the origin and insertion intact.
5. At the end of the laboratory period, wrap your specimen in cloth and store it in a tight, heavy-duty plastic bag. Discard the skin that was removed from your specimen and the plastic shipment bag. Wet your specimen from time to time with a preservative solution. Caution is necessary when using wetting solution, as it is caustic and poisonous if misused or used in a concentrated form.

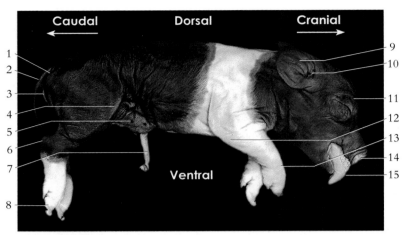

Figure 17.29 The directional terminology and superficial structures in a fetal pig (quadrupedal vertebrate).

1. Anus
2. Tail
3. Scrotum
4. Knee
5. Teat
6. Ankle
7. Umbilical cord
8. Hoof
9. Auricle (pinna)
10. External auditory canal
11. Superior palpebra (superior eyelid)
12. Elbow
13. Wrist
14. Naris (nostril)
15. Tongue

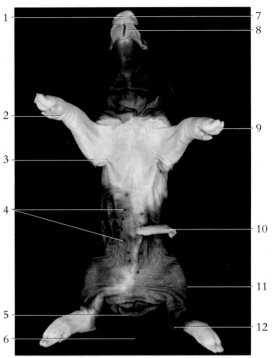

Figure 17.30 A ventral view of the surface anatomy of the fetal pig.

1. Nose
2. Wrist
3. Elbow
4. Teats
5. Scrotum
6. Tail
7. Nostril
8. Tongue
9. Digit
10. Umbilical cord
11. Knee
12. Ankle

Figure 17.31 A lateral view of superficial musculature of the fetal pig.

1. Tibialis anterior m.
2. Peroneus tarius m.
3. Peroneus longus m.
4. Gastrocnemius m.
5. Tensor fasciae latae m.
6. Biceps femoris m.
7. Gluteus superficialis m.
8. Gluteus medius m.
9. External abdominal oblique m.
10. Serratus ventralis m.
11. Pectoralis profundus m.
12. Latissimus dorsi m.
13. Trapezius m.
14. Triceps brachii m. (long head)
15. Triceps brachii m. (lateral head)
16. Deltoid m.
17. Supraspinatus m.
18. Omotransversarius m.
19. Cleidooccipitalis m.
20. Platysma m.
21. Brachialis m.
22. Extensor carpi radialis m.
23. Extensor digiti m.
24. Extensor digitorum communis m.
25. Ulnaris lateralis m.
26. Flexor digitorum profundus m.

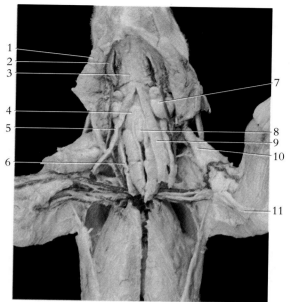

Figure 17.32 A ventral view of superficial muscles of the neck and upper torso.

1. Platysma m. (reflected)
2. Digastric m.
3. Mylohyoid m.
4. Sternohyoid m.
5. Omohyoid m.
6. Sternomastoid m.
7. Mandibular gland
8. Larynx
9. Sternothyroid m.
10. Brachiocephalic m.
11. Pectoralis superficialis m. (cut and reflected)

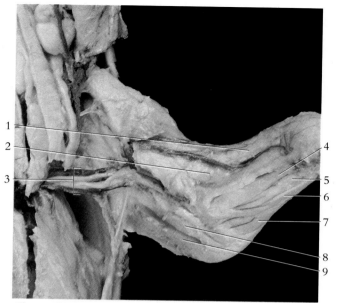

Figure 17.33 Superficial medial muscles of the forelimb.

1. Extensor carpi radialis m.
2. Biceps brachii m.
3. Axillary artery and vein, brachial plexus
4. Flexor carpi radialis m.
5. Flexor digitorum profundus m.
6. Flexor digitorum superficialis m.
7. Flexor carpi ulnaris m.
8. Triceps brachii m. (lateral head)
9. Triceps brachii m. (long head)

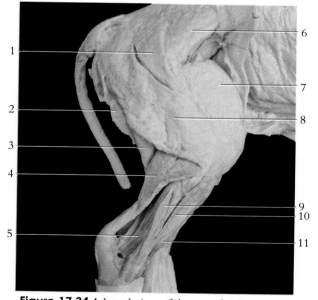

Figure 17.34 A lateral view of the superficial thigh and leg.

1. Gluteus superficialis m.
2. Semitendinosus m.
3. Semimembranosus m.
4. Gastrocnemius m.
5. Extensor digitorum quarti and quinti mm.
6. Gluteus medius m.
7. Tensor fasciae latae m.
8. Biceps femoris m.
9. Fibularis (peroneus) longus m.
10. Fibularis (peroneus) tertius m.
11. Tibialis anterior m.

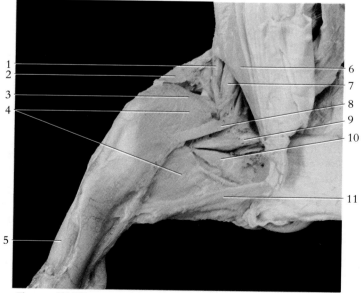

Figure 17.35 Medial muscles of the thigh and leg.

1. Iliacus m.
2. Tensor fasciae latae m.
3. Rectus femoris m.
4. Semimembranosus m.
5. Tibialis anterior m.
6. External abdominal oblique m.
7. Psoas major m.
8. Sartorius m.
9. Pectineus m.
10. Adductor m.
11. Semitendinosus m.

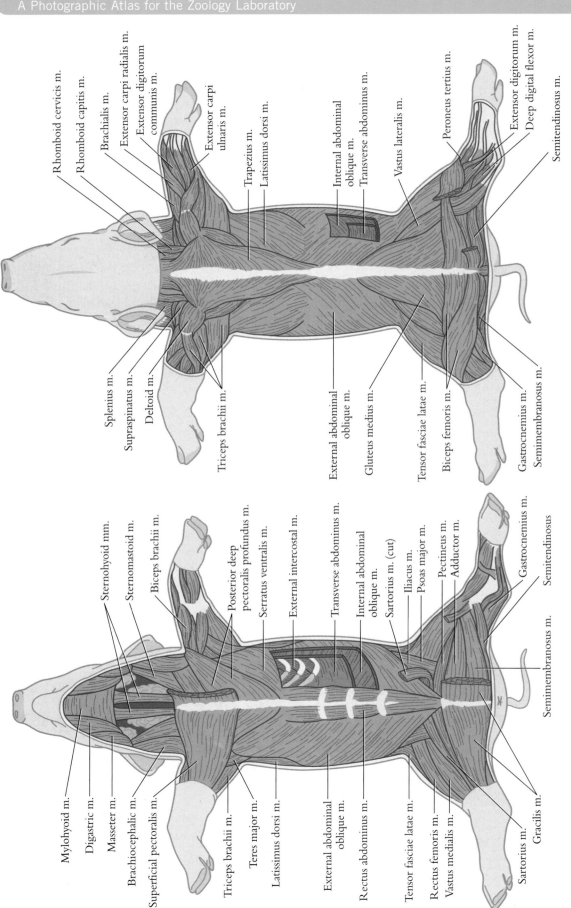

Figure 17.37 A dorsal view of the muscles of the fetal pig.

Figure 17.36 A ventral view of the muscles of the fetal pig.

Mammalia 169

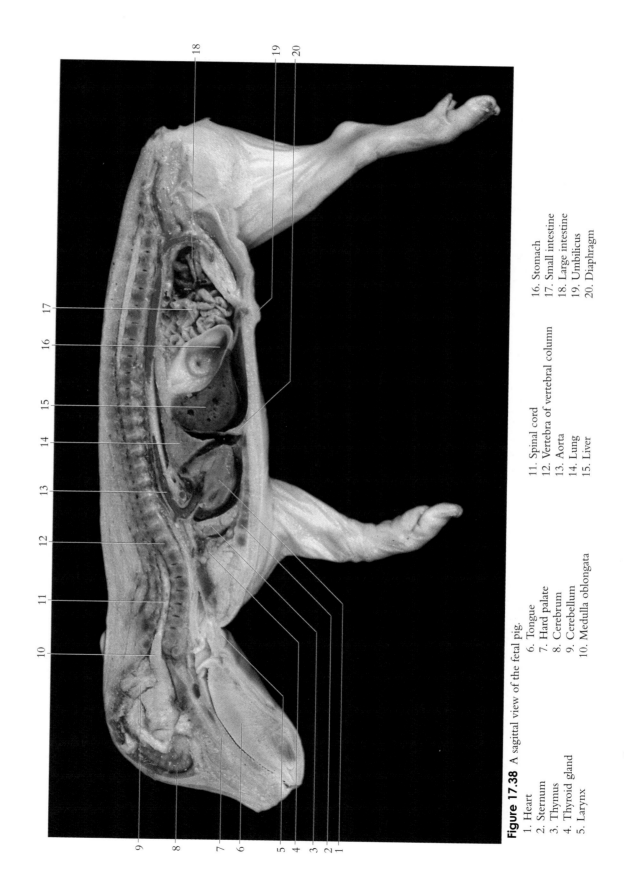

Figure 17.38 A sagittal view of the fetal pig.

1. Heart
2. Sternum
3. Thymus
4. Thyroid gland
5. Larynx
6. Tongue
7. Hard palate
8. Cerebrum
9. Cerebellum
10. Medulla oblongata
11. Spinal cord
12. Vertebra of vertebral column
13. Aorta
14. Lung
15. Liver
16. Stomach
17. Small intestine
18. Large intestine
19. Umbilicus
20. Diaphragm

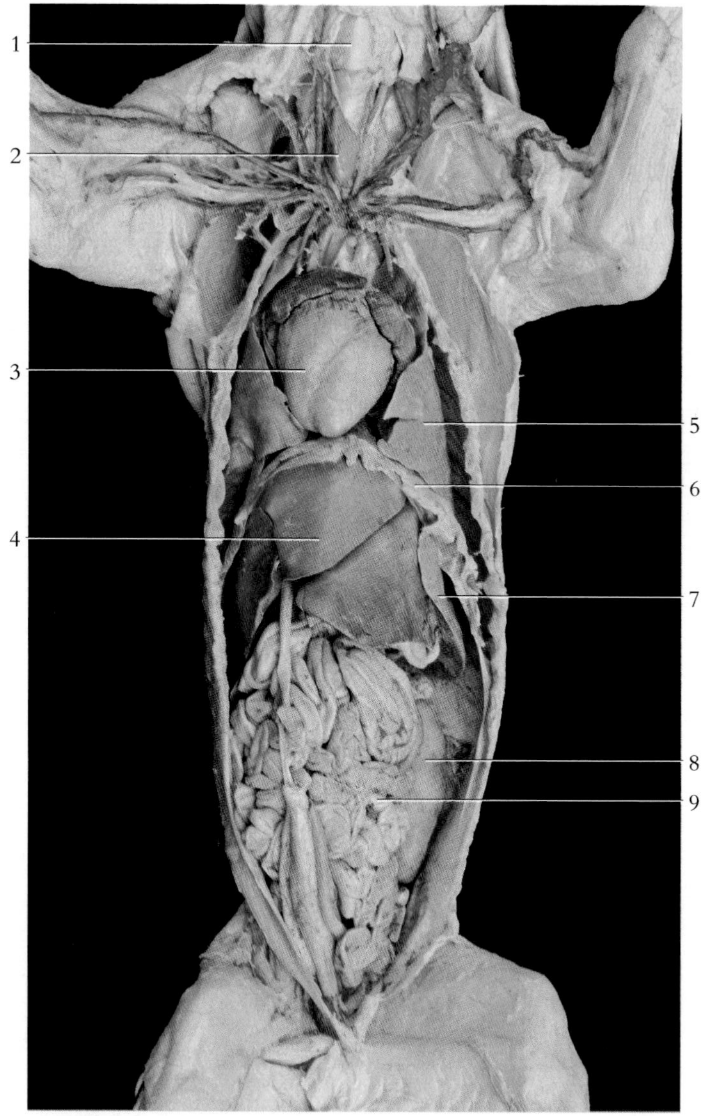

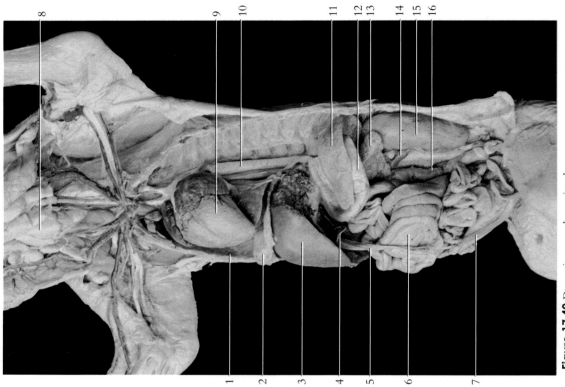

Figure 17.39 A ventral view of the viscera of a fetal pig.

1. Larynx
2. Thyroid gland
3. Heart
4. Liver
5. Lung
6. Diaphragm
7. Spleen
8. Kidney
9. Small intestine

Figure 17.40 Deep viscera and associated structures.

1. Lung
2. Diaphragm
3. Liver (cut)
4. Gallbladder
5. Umbilical vein
6. Small intestine
7. Umbilical artery
8. Larynx
9. Heart
10. Thoracic aorta
11. Spleen
12. Stomach
13. Pancreas
14. Renal vein
15. Kidney
16. Colon

Mammalia

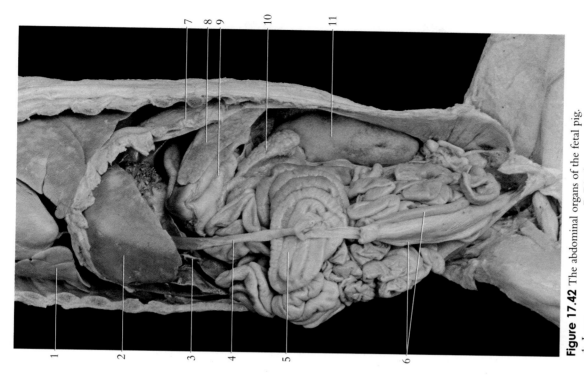

Figure 17.42 The abdominal organs of the fetal pig.
1. Lung
2. Liver (cut)
3. Gallbladder
4. Umbilical vein
5. Small intestine
6. Umbilical arteries
7. Diaphragm
8. Spleen
9. Stomach
10. Pancreas
11. Kidney

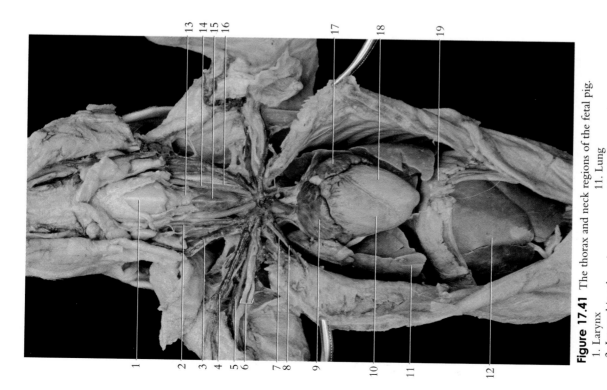

Figure 17.41 The thorax and neck regions of the fetal pig.
1. Larynx
2. Internal jugular vein
3. External jugular vein
4. Subclavian artery
5. Subclavian vein
6. Brachial plexus
7. Internal thoracic artery
8. Internal thoracic vein
9. Right auricle
10. Right ventricle
11. Lung
12. Liver
13. Trachea
14. Left common carotid artery
15. Thyroid gland
16. Vagus nerve
17. Left auricle
18. Left ventricle
19. Diaphragm

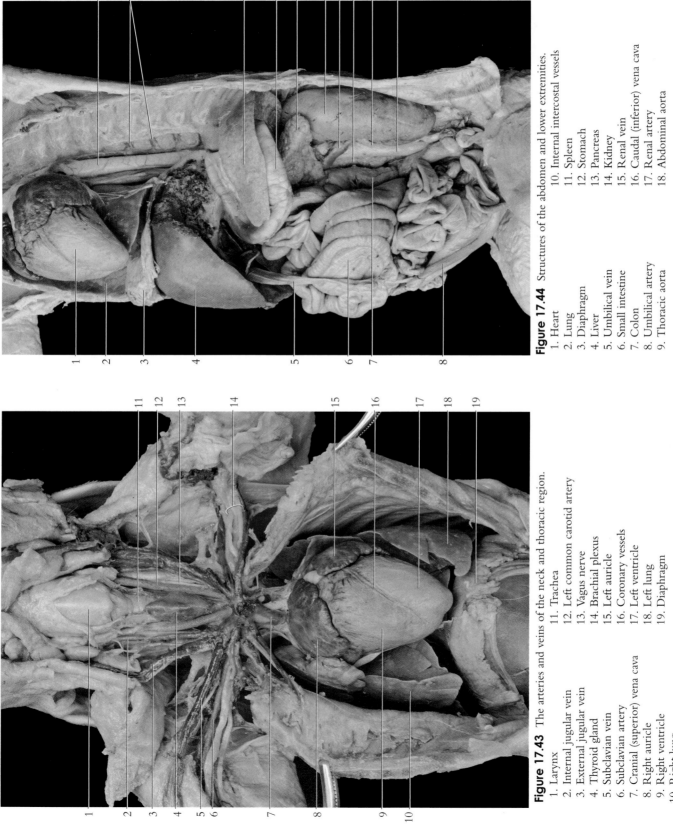

Figure 17.43 The arteries and veins of the neck and thoracic region.

1. Larynx
2. Internal jugular vein
3. External jugular vein
4. Thyroid gland
5. Subclavian vein
6. Subclavian artery
7. Cranial (superior) vena cava
8. Right auricle
9. Right ventricle
10. Right lung
11. Trachea
12. Left common carotid artery
13. Vagus nerve
14. Brachial plexus
15. Left auricle
16. Coronary vessels
17. Left ventricle
18. Left lung
19. Diaphragm

Figure 17.44 Structures of the abdomen and lower extremities.

1. Heart
2. Lung
3. Diaphragm
4. Liver
5. Umbilical vein
6. Small intestine
7. Colon
8. Umbilical artery
9. Thoracic aorta
10. Internal intercostal vessels
11. Spleen
12. Stomach
13. Pancreas
14. Kidney
15. Renal vein
16. Caudal (inferior) vena cava
17. Renal artery
18. Abdominal aorta

Mammalia

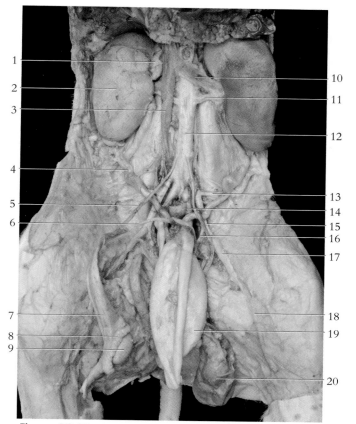

Figure 17.45 The urogenital system of the fetal pig.
1. Adrenal gland
2. Right kidney
3. Caudal (inferior) vena cava
4. Ureter
5. Genital vessels
6. Ductus (vas) deferens
7. Spermatic cord
8. Epididymis
9. Testis
10. Renal vein
11. Renal artery
12. Descending aorta
13. Iliolumbar artery
14. Rectum (cut)
15. Common iliac artery
16. Internal iliac artery
17. External iliac artery
18. Femoral artery
19. Urinary bladder
20. Testis

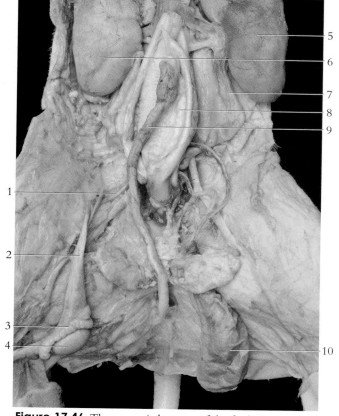

Figure 17.46 The urogenital system of the fetal pig.
1. Vas (ductus) deferens
2. Spermatic cord
3. Epididymis
4. Right testis
5. Left kidney
6. Right kidney
7. Ureter
8. Urinary bladder
9. Penis
10. Left testis

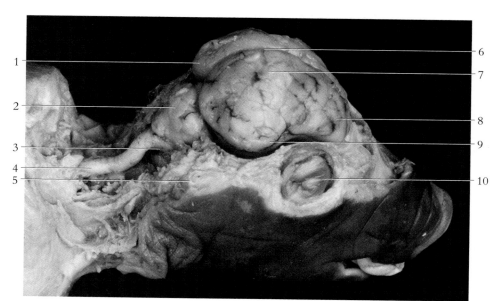

Figure 17.47 The general structures of the fetal pig brain. Because the cerebrum is less defined in pigs, the regions are not known as lobes as they are in humans.
1. Occipital region of cerebrum
2. Cerebellum
3. Medulla oblongata
4. Spinal cord
5. External acoustic meatus
6. Longitudinal fissure
7. Parietal region of cerebrum
8. Frontal region of cerebrum
9. Temporal region of cerebrum
10. Eye

Rabbit Dissection

Embalmed rabbits are often used in learning basic mammalian anatomy. Before the muscles and viscera can be studied, the specimen's skin has to be removed according to the following suggested guidelines.

1. Place the rabbit on a dissecting tray dorsal side up. Using a sharp scalpel, make a short, shallow incision through the skin across the nape of the neck. With your scissors, continue a dorsal midline incision forward over the skull and down the back to the tail.
2. Make a shallow incision around the neck and down each front leg to the paws. Continue a circular incision around each wrist. Beginning at the base of the tail, make incisions down each of the hind legs to the ankles. Make a circular cut around each ankle, and around the tail.
3. Carefully remove the skin, using your fingers or a blunt probe to separate the skin from underlying connective tissue. Where it is necessary to use a scalpel, keep the cutting edge directed toward the skin away from the muscle. If your specimen is a male, make an incision around the genitalia, leaving the skin intact. If your specimen is a female, the mammary glands will appear as longitudinal, glandular masses along the ventral side of the abdomen and thorax. They should be removed with the skin.
4. After the specimen is skinned, remove the excess fat and connective tissue to expose the underlying muscles. Make certain that the muscles are separated along their natural boundaries. If a transection of a muscle is necessary, isolate the muscle from its attached connective tissue, and make a clean cut across the belly of the muscle, leaving the origin and insertion intact.
5. At the end of the laboratory period, wrap your specimen in cloth and store it in a tight, heavy-duty plastic bag. Wet your specimen from time to time with a preservative solution. Caution has to be taken when using a phenol wetting solution, as it is caustic and poisonous if misused or used in a concentrated form.

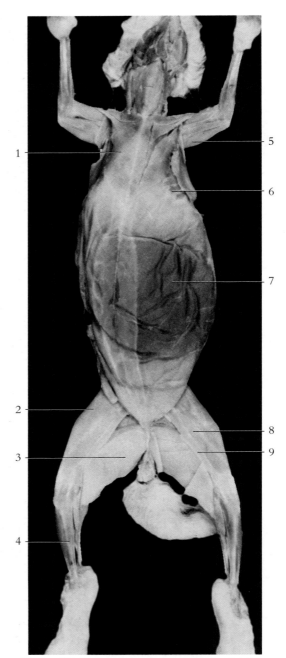

Figure 17.48 A ventral view of the superficial musculature of the rabbit. (m. = muscle)

1. Pectoralis major m.
2. Tensor fasciae latae m.
3. Gracilis m.
4. Tibialis anterior m.
5. Triceps brachii m. (long head)
6. Serratus ventralis m.
7. External abdominal oblique m.
8. Vastus medialis m.
9. Sartorius m.

Mammalia 175

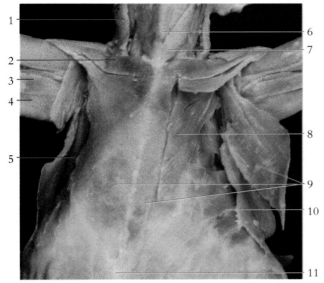

Figure 17.49 The neck and thoracic musculature of the rabbit.
1. Basioclavicularis m.
2. Pectoralis tenius m.
3. Epitrochlearis m.
4. Triceps brachii m. (long head)
5. Teres major m.
6. Sternohyoid m.
7. Sternomastoid m.
8. Pectoralis minor m.
9. Pectoralis major m.
10. Serratus ventralis m.
11. Linea alba

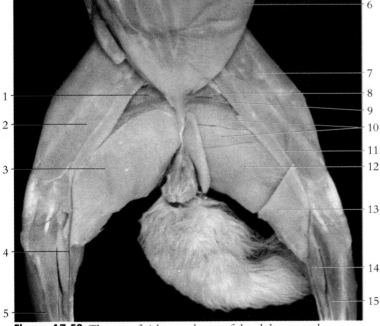

Figure 17.50 The superficial musculature of the abdomen and hindlimbs of the rabbit.
1. Pectineus m.
2. Vastus medialis m.
3. Gracilis m.
4. Gastrocnemius m.
5. Tibialis anterior m.
6. External abdominal oblique m.
7. Rectus femoris m.
8. Pectineus m.
9. Adductor brevis m.
10. Adductor longus m.
11. Sartorius m.
12. Semimembranosus m.
13. Gracilis m. (cut)
14. Soleus m.
15. Flexor digitorum longus m.

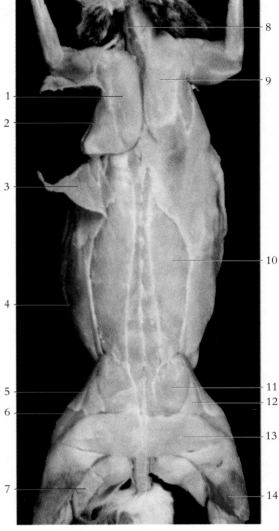

Figure 17.51 A dorsal view of the superficial musculature of the rabbit.
1. Supraspinatus m.
2. Infraspinatus m.
3. Spinotrapezius m. (cut)
4. External abdominal oblique m.
5. Tensor fasciae latae m.
6. Vastus lateralis m.
7. Semimembranosus proprius m.
8. Rhomboideus capitis m.
9. Acromiotrapezius m.
10. Longissimus dorsi m.
11. Gluteus medius m.
12. Caudofemoralis m.
13. Biceps femoris m. (cranial portion)
14. Biceps femoris m. (caudal portion)

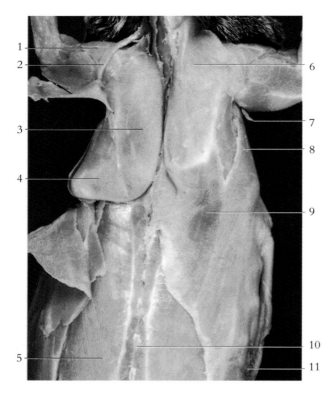

Figure 17.52 A dorsal view of the musculature of the shoulder thoracic and abdominal regions of the rabbit.

1. Cleidodeltoid m.
2. Levator scapulae ventralis m.
3. Supraspinatus m.
4. Infraspinatus m.
5. Longissimus dorsi m.
6. Acromiotrapezius m.
7. Cutaneous maximus m. (cut)
8. Latissimus dorsi m.
9. Spinotrapezius m.
10. Multifidus m.
11. External abdominal oblique m.

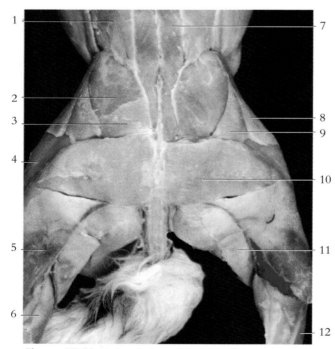

Figure 17.53 A dorsal view of the musculature of the hip and hindlimbs of the rabbit.

1. Longissimus dorsi m.
2. Gluteus medius m.
3. Gluteus maximus m.
4. Vastus lateralis m.
5. Biceps femoris m. (caudal portion)
6. Gastrocnemius m.
7. Multifidus m.
8. Tensor fasciae latae m.
9. Caudofemoralis m.
10. Biceps femoris m. (cranial portion)
11. Semimembranosus m.
12. Tibialis anterior m.

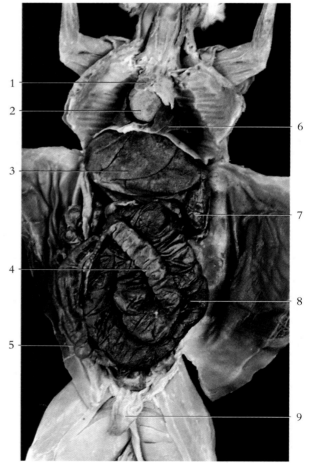

Figure 17.54 A ventral view of the rabbit viscera.

1. Thymus gland
2. Heart
3. Liver
4. Ileum of small intestine
5. Caecum
6. Left lung
7. Stomach
8. Colon
9. Urinary bladder

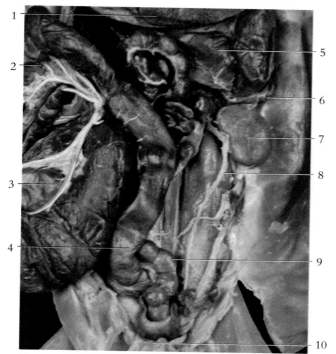

Figure 17.55 A ventral view of the abdominal viscera of the rabbit.
1. Liver
2. Caecum
3. Sacculus rotundus
4. Colon
5. Stomach
6. Spleen
7. Kidney
8. Ureter
9. Descending colon
10. Urinary bladder

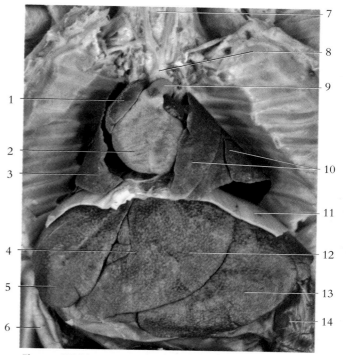

Figure 17.56 A ventral view of the viscera of the thoracic and upper abdominal cavities of the rabbit.
1. Right atrium of heart
2. Right ventricle of heart
3. Right lung
4. Quadrate lobe of liver
5. Right lobe of liver
6. Duodenum
7. Trachea
8. Aortic arch
9. Pulmonary trunk
10. Left lung
11. Diaphragm
12. Left median lobe of liver
13. Left lateral lobe of liver
14. Stomach

Cat Dissection

Embalmed cats purchased from biological supply houses are excellent specimens for dissecting and learning basic mammalian anatomy. Before the muscles and viscera of a cat can be studied, the specimen's skin has to be removed according to the following suggested guidelines.

1. Place the cat on a dissecting tray dorsal side up. Using a sharp scalpel, make a short, shallow incision through the skin across the nape of the neck. With your scissors, continue a dorsal midline incision forward over the skull and down the back to about two inches onto the tail. Sever the tail with bone shears or a saw and discard.
2. Make a shallow incision around the neck and down each front leg to the paws. Continue a circular incision around each wrist. Beginning at the base of the tail, make incisions down each of the hind legs to the ankles. Make a circular cut around each ankle.
3. Carefully remove the skin, using your fingers or a blunt probe to separate the skin from underlying connective tissue. Where it is necessary to use a scalpel, keep the cutting edge directed toward the skin away from the muscle. If your specimen is a male, make an incision around the genitalia, leaving the skin intact. If your specimen is a female, the mammary glands will appear as longitudinal, glandular masses along the ventral side of the abdomen and thorax. They may be removed with the skin.
4. After the specimen is skinned, remove the excess fat and connective tissue to expose the underlying muscles. Make certain that the muscles are separated along their natural boundaries. If a transection of a muscle is necessary, isolate the muscle from its attached connective tissue, and make a clean cut across the belly of the muscle, leaving the origin and insertion intact.
5. At the end of the laboratory period, wrap your specimen in cloth and store it in a tight, heavy-duty plastic bag. Wet your specimen from time to time with a preservative solution. Caution has to be taken when using a phenol wetting solution, as it is caustic and poisonous if misused or used in a concentrated form.

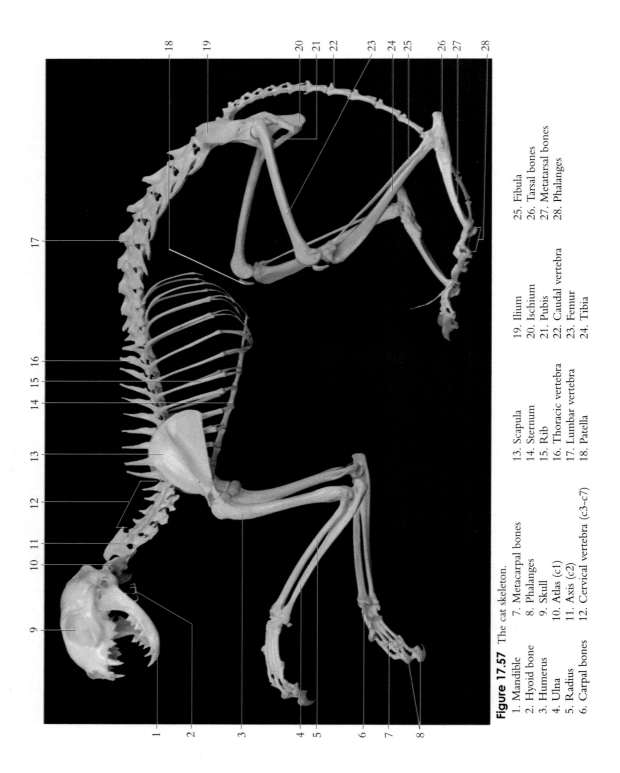

Figure 17.57 The cat skeleton.

1. Mandible
2. Hyoid bone
3. Humerus
4. Ulna
5. Radius
6. Carpal bones
7. Metacarpal bones
8. Phalanges
9. Skull
10. Atlas (c1)
11. Axis (c2)
12. Cervical vertebra (c3–c7)
13. Scapula
14. Sternum
15. Rib
16. Thoracic vertebra
17. Lumbar vertebra
18. Patella
19. Ilium
20. Ischium
21. Pubis
22. Caudal vertebra
23. Femur
24. Tibia
25. Fibula
26. Tarsal bones
27. Metatarsal bones
28. Phalanges

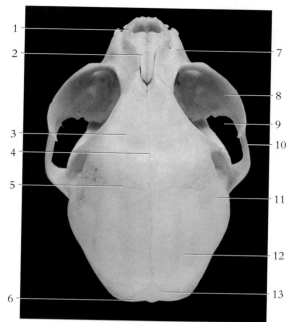

Figure 17.58 A dorsal view of a cat skull.
1. Premaxilla
2. Nasal bone
3. Frontal bone
4. Sagittal suture
5. Coronal suture
6. Nuchal crest
7. Maxilla
8. Zygomatic (malar) bone
9. Orbit
10. Zygomatic arch
11. Temporal bone
12. Parietal bone
13. Interparietal bone

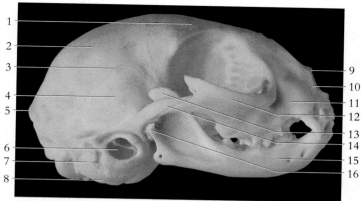

Figure 17.59 A lateral view of a cat skull.
1. Frontal bone
2. Parietal bone
3. Squamosal suture
4. Temporal bone
5. Nuchal crest
6. External acoustic meatus
7. Mastoid process
8. Tympanic bulla
9. Nasal bone
10. Premaxilla bone
11. Maxilla
12. Zygomatic (malar) bone
13. Coronoid process of mandible
14. Zygomatic arch
15. Mandible
16. Condylar process of mandible

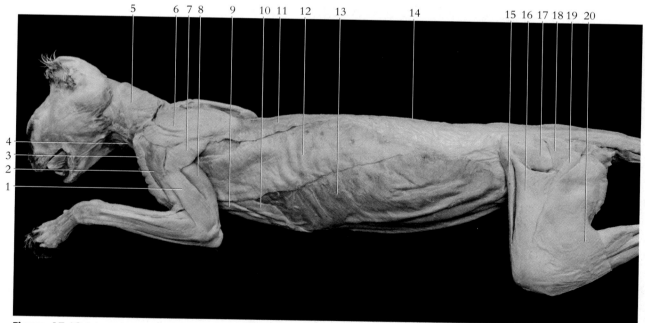

Figure 17.60 A lateral view of the superficial muscles of the cat.
1. Lateral head of triceps brachii m.
2. Acromiodeltoid m.
3. Clavobrachialis m. (clavodeltiod)
4. Sternomastoid m.
5. Clavotrapezius m.
6. Acromiotrapezius m.
7. Spinodeltoid m.
8. Long head of triceps brachii m.
9. Pectoralis minor m.
10. Xiphihumeralis m.
11. Spinotrapezius m.
12. Latissimus dorsi m.
13. External abdominal oblique m.
14. Lumbodorsal fascia
15. Sartorius m.
16. Tensor fasciae latae m.
17. Gluteus medius m.
18. Gluteus maximus m.
19. Caudofemoralis m.
20. Biceps femoris m.

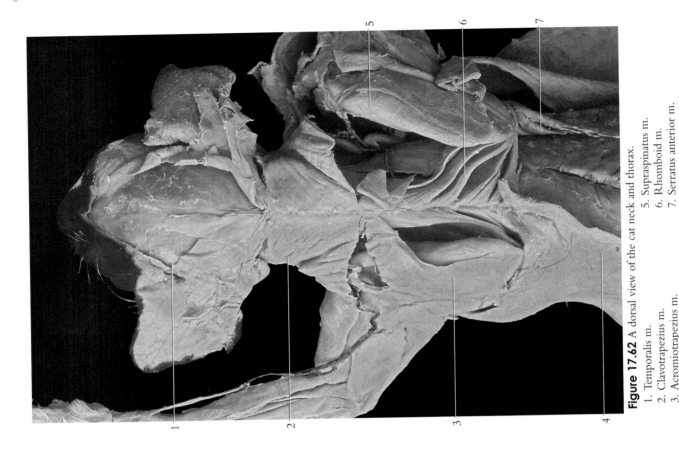

Figure 17.62 A dorsal view of the cat neck and thorax.
1. Temporalis m.
2. Clavotrapezius m.
3. Acromiotrapezius m.
4. Latissimus dorsi m.
5. Supraspinatus m.
6. Rhomboid m.
7. Serratus anterior m.

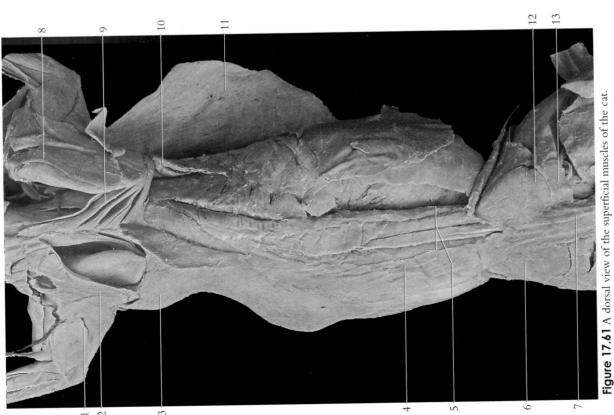

Figure 17.61 A dorsal view of the superficial muscles of the cat.
1. Lateral head of triceps brachii m.
2. Acromiotrapezius m.
3. Latissimus dorsi m.
4. Lumbodorsal fascia
5. Sacrospinalis m.
6. Gluteus medius m.
7. Caudal m.
8. Supraspinatus m.
9. Rhomboid m.
10. Serratus anterior m.
11. Latissimus dorsi m.
12. Gluteus maximus m.
13. Caudofemoralis m.

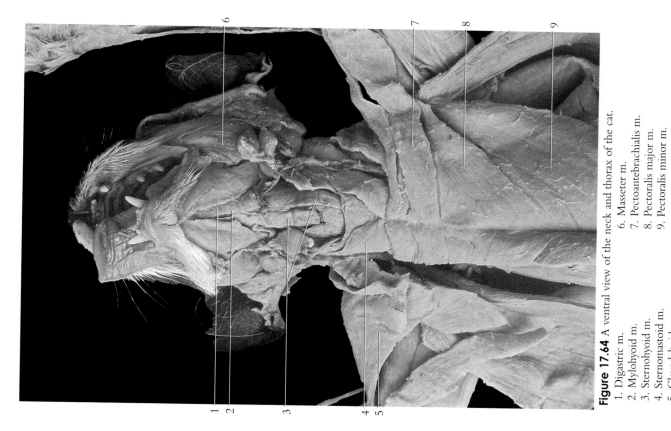

Figure 17.64 A ventral view of the neck and thorax of the cat.

1. Digastric m.
2. Mylohyoid m.
3. Sternohyoid m.
4. Sternomastoid m.
5. Clavodeltoid m.
6. Masseter m.
7. Pectoantebrachialis m.
8. Pectoralis major m.
9. Pectoralis minor m.

Figure 17.63 A superficial view of the ventral trunk of the female cat.

1. Mammary glands
2. Nipples
3. External abdominal oblique m.

Mammalia 181

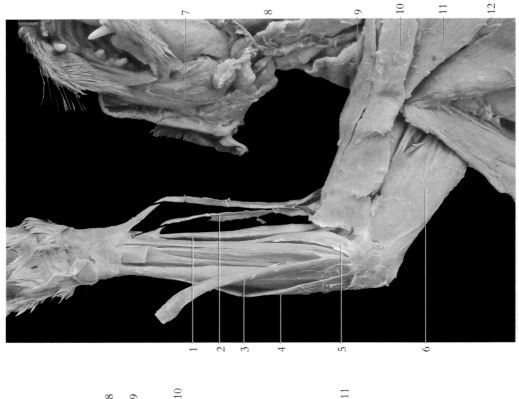

Figure 17.65 A lateral view of the shoulder and brachium of the cat.

1. Acromiotrapezius m.
2. Levator scapulae ventralis m.
3. Spinodeltoid m.
4. Latissimus dorsi m.
5. Long head of triceps brachii m.
6. Clavobrachialis m. (clavodeltoid)
7. Lateral head of triceps brachii m.
8. Clavotrapezius m.
9. Parotid gland
10. Acromiodeltoid m.
11. Brachioradialis m.

Figure 17.66 An anterior view of the brachium and antebrachium of the cat.

1. Extensor carpi radialis longus m.
2. Brachioradialis m.
3. Palmaris longus m. (cut)
4. Flexor carpi ulnaris m.
5. Pronator teres m.
6. Epitrochlearis
7. Masseter m.
8. Sternomastoid m.
9. Clavobrachialis m. (clavodeltiod)
10. Pectoantebrachialis m.
11. Pectoralis major m.
12. Pectoralis minor m.

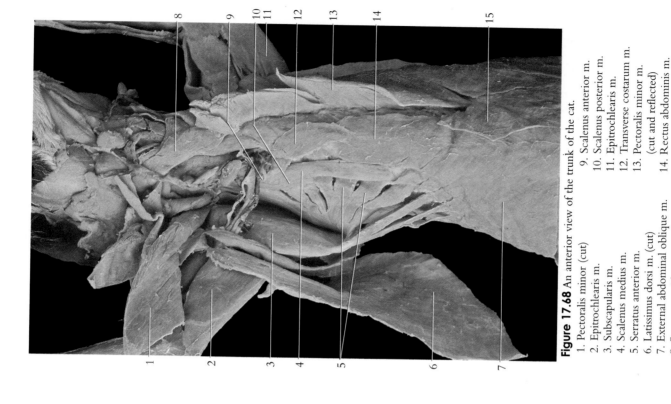

Figure 17.68 An anterior view of the trunk of the cat.

1. Pectoralis minor (cut)
2. Epitrochlearis m.
3. Subscapularis m.
4. Scalenus medius m.
5. Serratus anterior m.
6. Latissimus dorsi m. (cut)
7. External abdominal oblique m.
8. Sternomastoid m.
9. Scalenus anterior m.
10. Scalenus posterior m.
11. Epitrochlearis m.
12. Transverse costarum m.
13. Pectoralis minor m. (cut and reflected)
14. Rectus abdominis m.
15. Xiphihumeralis m. (cut)

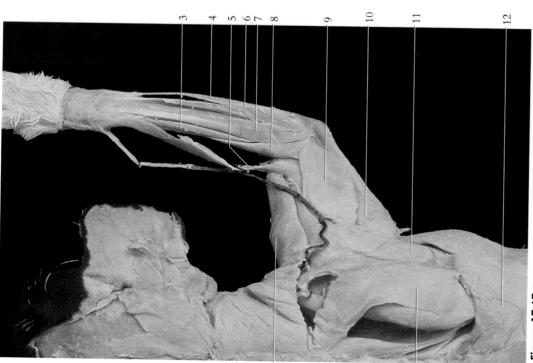

Figure 17.67 A posterior view of the brachium and antebrachium of the cat.

1. Clavobrachialis m.
2. Acromiotrapezius m.
3. Extensor carpi radialis brevis m.
4. Extensor carpi ulnaris m.
5. Brachioradialis m.
6. Extensor digitorum lateralis m.
7. Extensor digitorum communis m.
8. Extensor carpi radialis longus m.
9. Lateral head of triceps brachii m.
10. Long head of triceps brachii m.
11. Spinodeltoid m.
12. Latissimus dorsi m.

Figure 17.69 A lateral view of the trunk of the cat.
1. Internal abdominal oblique m.
2. Tensor fascia latae
3. Caudofemoralis m.
4. Vastus lateralis m.
5. Sartorius m.
6. External abdominal oblique m. (cut and reflected)
7. Latissimus dorsi m.
8. Spinodeltoid m.
9. Transverse abdominis m.
10. Serratus anterior m.
11. Long head of triceps brachii m.

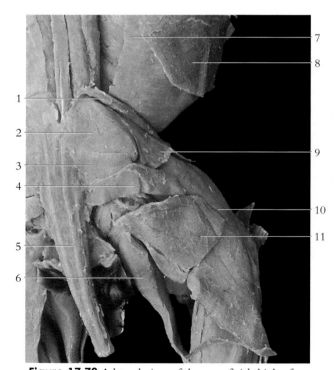

Figure 17.70 A lateral view of the superficial thigh of the cat.
1. Sartorius m.
2. Gluteus medius m.
3. Gluteus maximus m.
4. Caudofemoralis m.
5. Caudal m.
6. Semitendinosus m.
7. Internal abdominal oblique m.
8. External abdominal oblique m.
9. Tensor fascia latae (cut)
10. Vastus lateralis m.
11. Biceps femoris m.

Figure 17.71 A medial view of the thigh and leg of the cat.
1. Sartorius m. (cut)
2. Vastus lateralis m.
3. Rectus femoris m.
4. Vastus medialis m.
5. Flexor digitorum longus m.
6. Tibialis anterior m.
7. Rectus abdominis m.
8. Adductor longus m.
9. Adductor femoris m.
10. Semimembranosus m.
11. Gracilis m. (cut)
12. Gastrocnemius m.
13. Tendo calcaneus (Achilles' tendon)

Mammalia 185

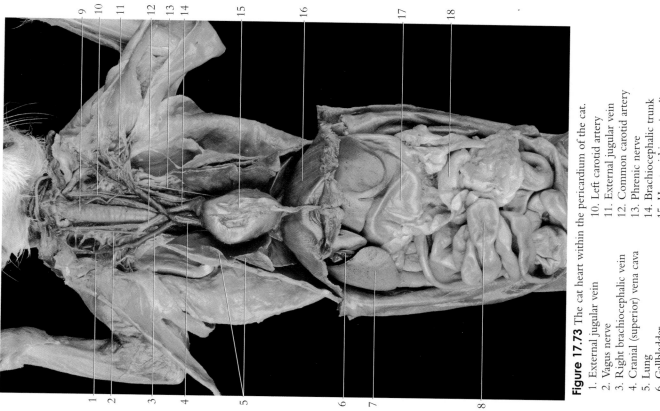

Figure 17.73 The cat heart within the pericardium of the cat.

1. External jugular vein
2. Vagus nerve
3. Right brachiocephalic vein
4. Cranial (superior) vena cava
5. Lung
6. Gallbladder
7. Liver
8. Small intestine
9. Trachea
10. Left carotid artery
11. External jugular vein
12. Common carotid artery
13. Phrenic nerve
14. Brachiocephalic trunk
15. Heart within pericardium
16. Diaphragm
17. Stomach
18. Greater omentum

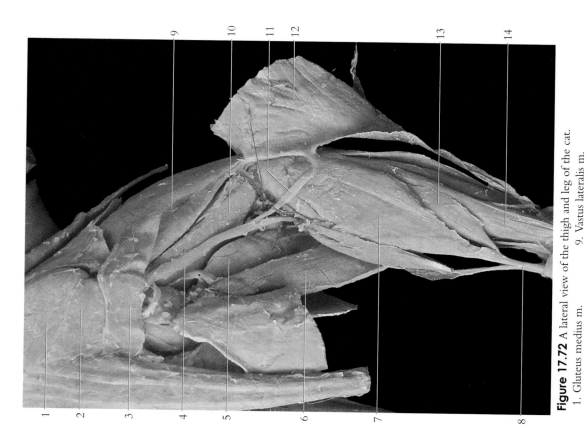

Figure 17.72 A lateral view of the thigh and leg of the cat.

1. Gluteus medius m.
2. Gluteus maximus m.
3. Caudofemoralis m.
4. Sciatic nerve
5. Semimembranosus m.
6. Semitendinosus m.
7. Gastrocnemius m.
8. Tendo calcaneus
9. Vastus lateralis m.
10. Adductor femoris m.
11. Tenuissimus m.
12. Biceps femoris m. (cut and reflected)
13. Soleus m.
14. Fibularis (peroneal) m.

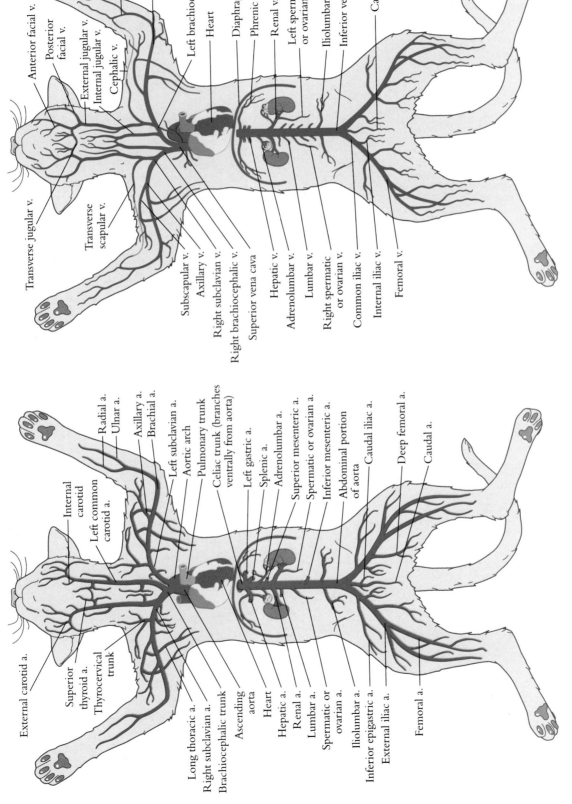

Figure 17.75 The principal veins of the cat, ventral view (v = vein).

Figure 17.74 The principal arteries of the cat, ventral view. (a = artery).

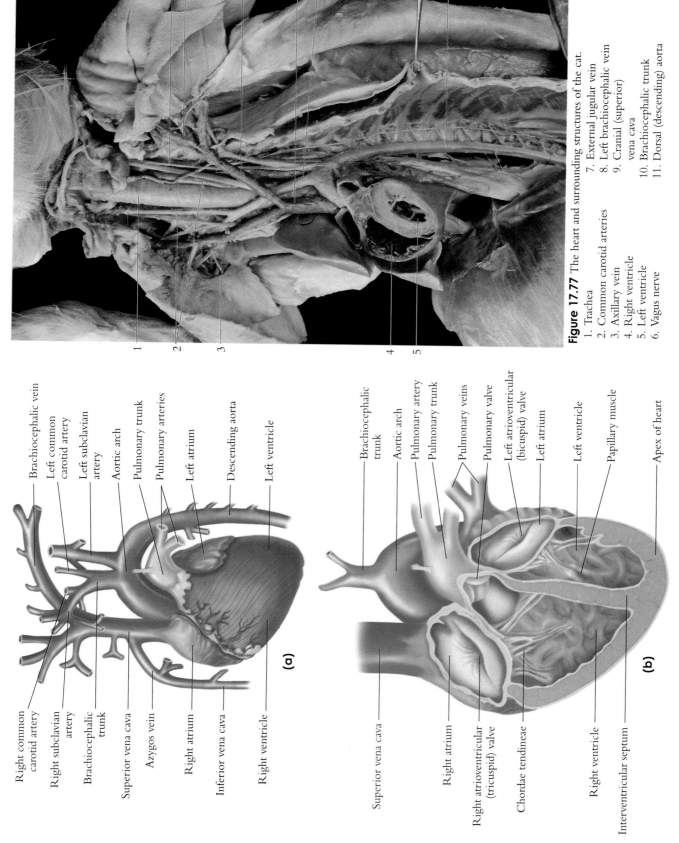

Figure 17.77 The heart and surrounding structures of the cat.
1. Trachea
2. Common carotid arteries
3. Axillary vein
4. Right ventricle
5. Left ventricle
6. Vagus nerve
7. External jugular vein
8. Left brachiocephalic vein
9. Cranial (superior) vena cava
10. Brachiocephalic trunk
11. Dorsal (descending) aorta

Figure 17.76 The cat heart and associated vessels (a) and the internal anatomy (b).

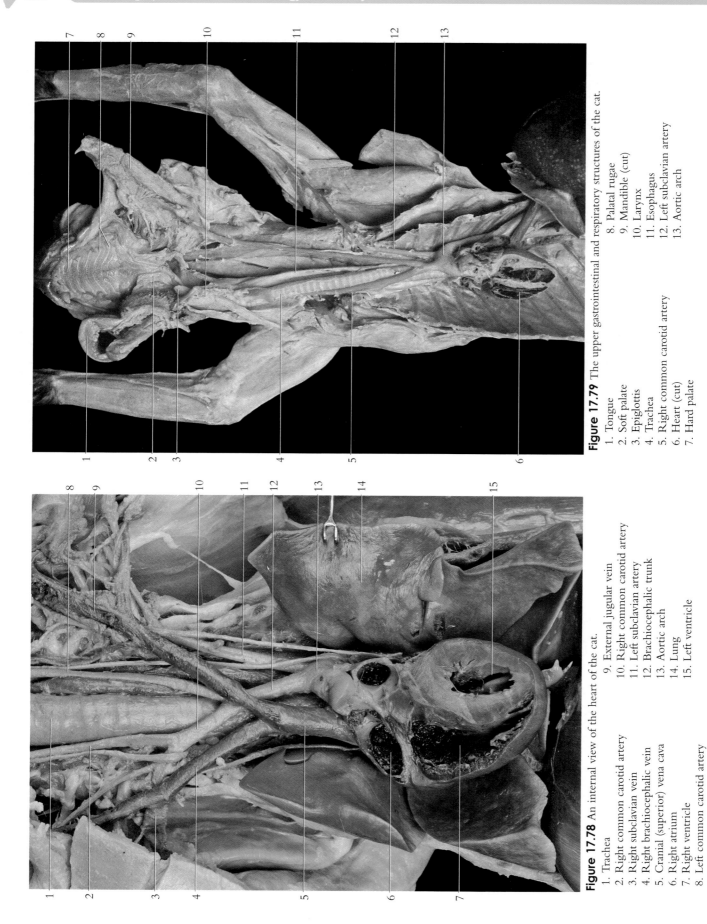

Figure 17.78 An internal view of the heart of the cat.
1. Trachea
2. Right common carotid artery
3. Right subclavian vein
4. Right brachiocephalic vein
5. Cranial (superior) vena cava
6. Right atrium
7. Right ventricle
8. Left common carotid artery
9. External jugular vein
10. Right common carotid artery
11. Left subclavian artery
12. Brachiocephalic trunk
13. Aortic arch
14. Lung
15. Left ventricle

Figure 17.79 The upper gastrointestinal and respiratory structures of the cat.
1. Tongue
2. Soft palate
3. Epiglottis
4. Trachea
5. Right common carotid artery
6. Heart (cut)
7. Hard palate
8. Palatal rugae
9. Mandible (cut)
10. Larynx
11. Esophagus
12. Left subclavian artery
13. Aortic arch

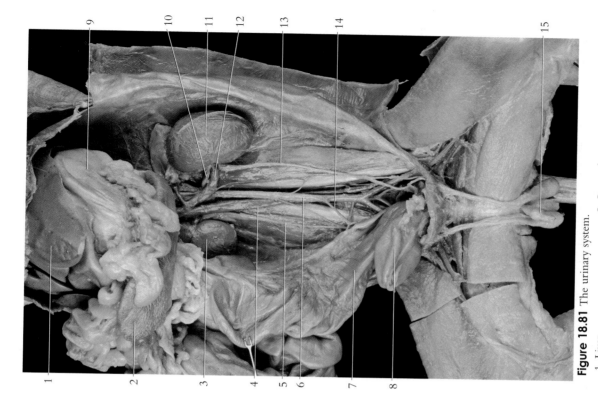

Figure 18.81 The urinary system.

1. Liver
2. Spleen (reflected)
3. Right kidney
4. Caudal (inferior) vena cava
5. Right ureter
6. Abdominal aorta
7. Colon
8. Urinary bladder
9. Stomach
10. Renal artery
11. Left kidney
12. Renal vein
13. Left ureter
14. Inferior mesenteric artery
15. Penis

Figure 18.80 An anterior view of the deep structures of the trunk of the cat.

1. Right common carotid artery
2. Vagus nerve
3. Heart (cut)
4. Thoracic aorta
5. Liver
6. Stomach
7. Spleen
8. Small intestine
9. Colon
10. Left brachiocephalic vein
11. Cranial (superior) vena cava
12. Aortic arch
13. Intercostal artery
14. Celiac trunk
15. Superior mesenteric artery
16. Kidney
17. Urinary bladder

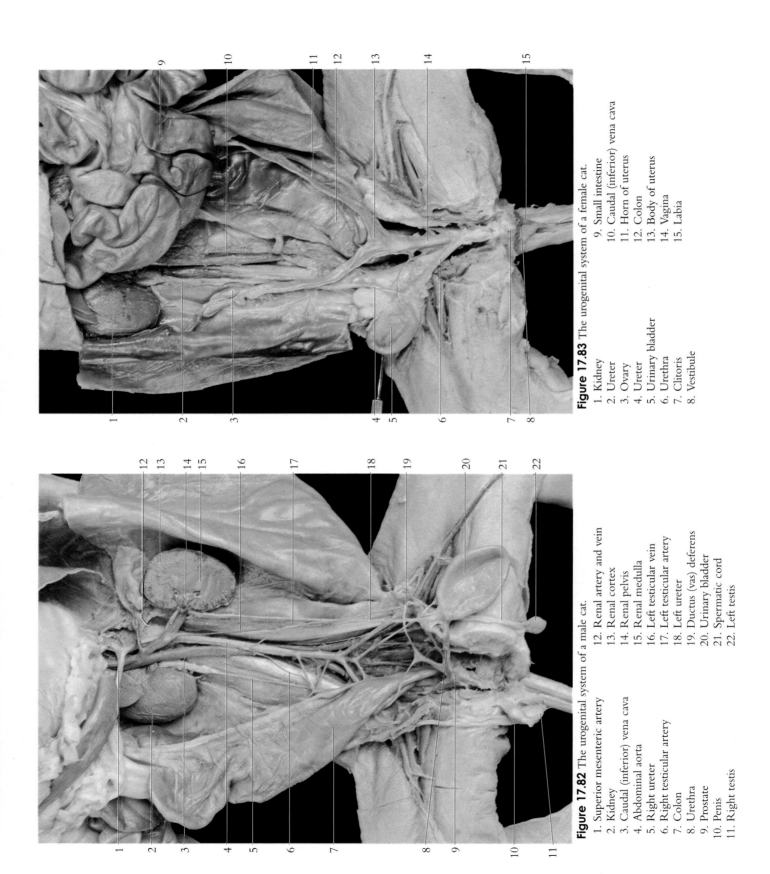

Figure 17.82 The urogenital system of a male cat.

1. Superior mesenteric artery
2. Kidney
3. Caudal (inferior) vena cava
4. Abdominal aorta
5. Right ureter
6. Right testicular artery
7. Colon
8. Urethra
9. Prostate
10. Penis
11. Right testis
12. Renal artery and vein
13. Renal cortex
14. Renal pelvis
15. Renal medulla
16. Left testicular vein
17. Left testicular artery
18. Left ureter
19. Ductus (vas) deferens
20. Urinary bladder
21. Spermatic cord
22. Left testis

Figure 17.83 The urogenital system of a female cat.

1. Kidney
2. Ureter
3. Ovary
4. Ureter
5. Urinary bladder
6. Urethra
7. Clitoris
8. Vestibule
9. Small intestine
10. Caudal (inferior) vena cava
11. Horn of uterus
12. Colon
13. Body of uterus
14. Vagina
15. Labia

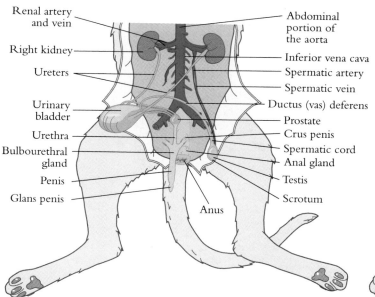

Figure 17.84 A diagram of the urogenital system of a male cat.

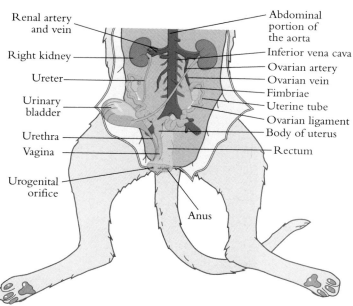

Figure 17.85 A diagram of the urogenital system of a female cat.

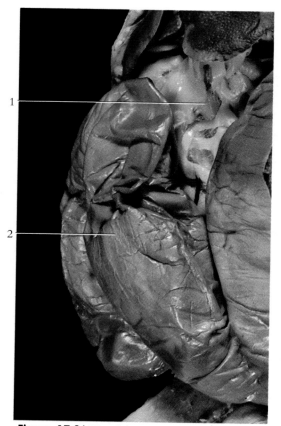

Figure 17.86 The abdominal cavity of a pregnant cat.
1. Greater omentum
2. Right horn of uterus

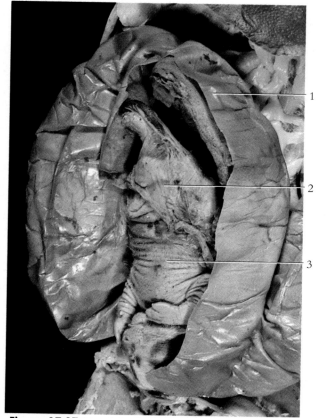

Figure 17.87 The abdominal cavity of a pregnant cat.
1. Uterine wall (cut)
2. Amniotic sac enclosing fetus
3. Fetus

Mammalian Heart and Brain Dissection

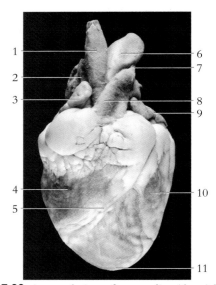

Figure 17.88 A ventral view of mammalian (sheep) heart.
1. Brachiocephalic artery
2. Cranial vena cava
3. Right auricle of right atrium
4. Right ventricle
5. Interventricular groove
6. Aortic arch
7. Ligamentum arteriosum
8. Pulmonary trunk
9. Left auricle of left atrium
10. Left ventricle
11. Apex of heart

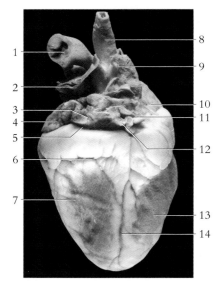

Figure 17.89 A dorsal view of mammalian (sheep) heart.
1. Aorta
2. Pulmonary artery
3. Pulmonary vein
4. Left auricle
5. Left atrium
6. Atrioventricular groove
7. Left ventricle
8. Brachiocephalic artery
9. Cranial vena cava
10. Right auricle
11. Right atrium
12. Pulmonary vein
13. Right ventricle
14. Interventricular groove

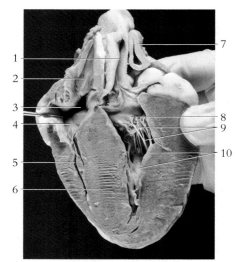

Figure 17.90 A coronal section of the mammalian (sheep) heart.
1. Aorta
2. Cranial vena cava
3. Right atrium
4. Right atrioventricular (tricuspid) valve
5. Right ventricle
6. Interventricular septum
7. Pulmonary artery
8. Left atrioventricular (bicuspid) valve
9. Chordae tendineae
10. Papillary muscles

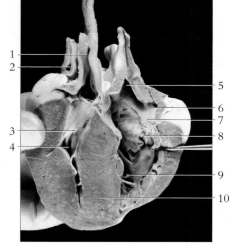

Figure 17.91 A coronal section of the mammalian (sheep) heart showing the valves.
1. Opening of the brachiocephalic artery
2. Pulmonary artery
3. Left atrioventricular (bicuspid) valve
4. Left ventricle
5. Opening of cranial vena cava
6. Opening of coronary sinus
7. Right atrium
8. Right atrioventricular (tricuspid) valve
9. Right ventricle
10. Interventricular septum

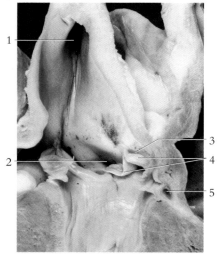

Figure 17.92 A coronal section of the mammalian (sheep) heart showing openings of coronary arteries.
1. Opening of brachiocephalic artery
2. Opening of left coronary artery
3. Opening of right coronary artery
4. Aortic valve
5. Coronary vessel

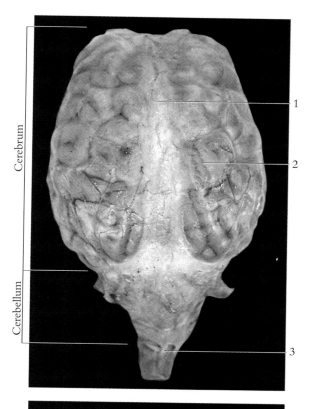

Figure 17.93 A dorsal view of the sheep brain.
1. Dura mater covering longitudinal cerebral fissure
2. Arachnoid
3. Medulla oblongata

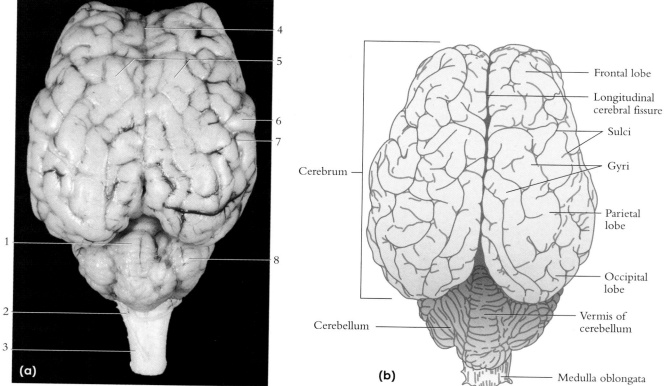

Figure 17.94 A dorsal view of the sheep brain, (a) photograph; (b) diagram.
1. Vermis of cerebellum
2. Medulla oblongata
3. Spinal cord
4. Longitudinal cerebral fissure
5. Cerebral hemispheres
6. Gyrus
7. Sulcus
8. Cerebellum

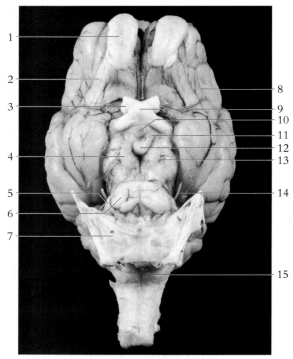

Figure 17.95 A ventral view of sheep brain with dura mater cut and reflected.
1. Olfactory bulb
2. Olfactory tract
3. Optic nerve
4. Oculomotor nerve
5. Trigeminal nerve
6. Pons
7. Dura mater (cut and reflected)
8. Pia mater (adhering to brain)
9. Optic chiasma
10. Position of pituitary stalk
11. Tuber cinereum
12. Mammillary body
13. Cerebral penduncle
14. Trochlear nerve
15. Medulla oblongata

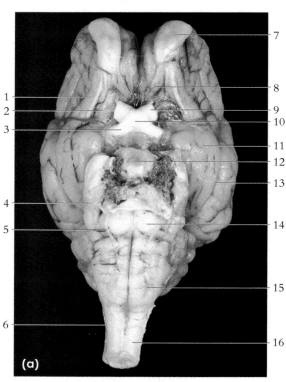

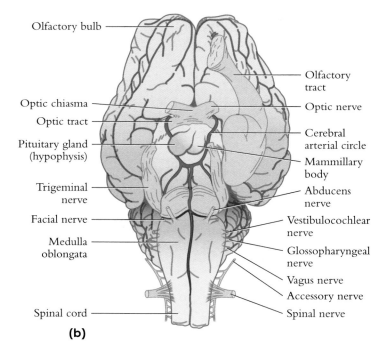

Figure 17.96 A ventral view of sheep brain, (a) photograph and (b) diagram.
1. Lateral olfactory band
2. Olfactory trigone
3. Optic tract
4. Trigeminal nerve
5. Abducens nerve
6. Accessory nerve
7. Olfactory bulb
8. Medial olfactory band
9. Optic nerve
10. Optic chiasma
11. Pyriform lobe
12. Pituitary gland (hypophysis)
13. Rhinal sulcus
14. Pons
15. Medulla oblongata
16. Spinal cord

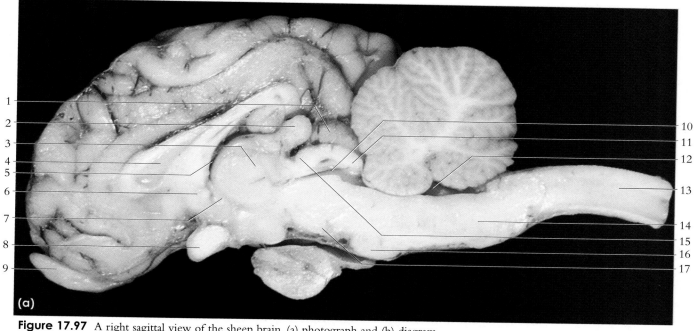

Figure 17.97 A right sagittal view of the sheep brain, (a) photograph and (b) diagram.

1. Superior colliculus
2. Pineal body (gland)
3. Intermediate mass of thalamus
4. Septum pellucidum
5. Interventricular foramen (foramen of Monro)
6. Anterior commissure
7. Third ventricle
8. Optic chiasma
9. Olfactory bulb
10. Mesencephalic (cerebral) aqueduct
11. Inferior colliculus
12. 4th ventricle
13. Spinal cord
14. Medulla oblongata
15. Posterior commissure
16. Pons
17. Cerebral peduncle

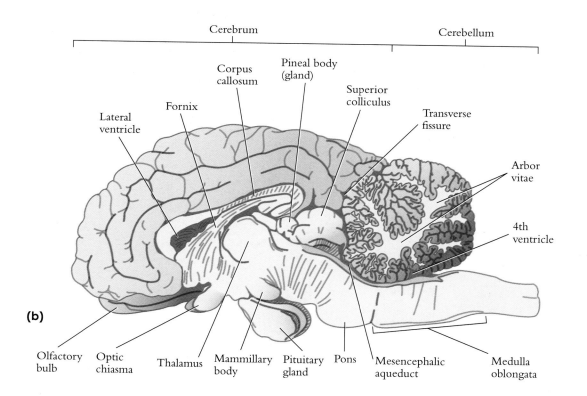

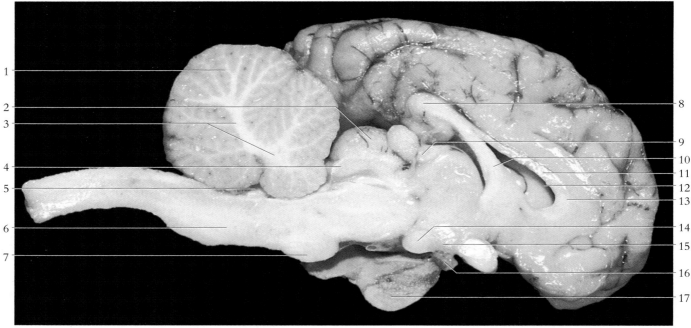

Figure 17.98 A left sagittal view of the sheep brain.
1. Cerebellum
2. Superior colliculus
3. Arbor vitae
4. Inferior colliculus
5. 4th ventricle
6. Medulla oblongata
7. Pons
8. Splenium of corpus callosum
9. Habenular trigone
10. Fornix
11. Body of corpus callosum
12. Lateral ventricle
13. Genu of corpus callosum
14. Mammillary body
15. Tuber cinereum of hypothalamus
16. Pituitary stalk
17. Pituitary gland (hypophysis)

Figure 17.99 A lateral view of the brainstem.
1. Pons
2. Abducens nerve
3. Medulla oblongata
4. Hypoglossal nerve
5. Spinal cord
6. Lateral geniculate body
7. Medial geniculate body
8. Trochlear nerve
9. Trigeminal nerve
10. Accessory nerve

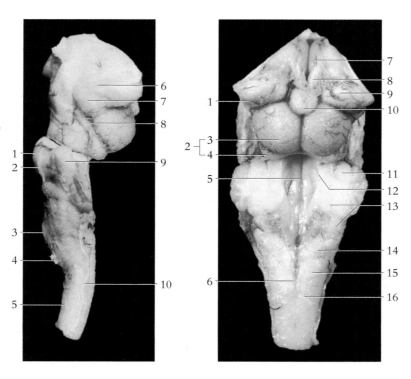

Figure 17.100 A dorsal view of the brainstem.
1. Medial geniculate body
2. Corpora quadrigemina
3. Superior colliculus
4. Inferior colliculus
5. 4th ventricle
6. Dorsal median sulcus
7. Intermediate mass of thalamus
8. Habenular trigone
9. Thalamus
10. Pineal gland
11. Middle cerebellar peduncle
12. Anterior cerebellar peduncle
13. Posterior cerebellar peduncle
14. Tuberculum cuneatum
15. Fasciculus gracilis
16. Fasciculus cuneatus

Human Biology

Chapter 18

Because humans are vertebrate organisms, the study of human biology is appropriate in a general zoology course. *Human anatomy* is the scientific discipline that investigates the structure of the body, and *human physiology* is the scientific discipline that investigates how body structures function. The purpose of this chapter is to present a visual overview of the principal anatomical structures of the human body.

Because both the *skeletal system* and the *muscular system* are concerned with body movement, they are frequently discussed together as the *skeletomusculature system*. In a functional sense, the flexible internal framework, or *bones* of the skeleton, support and provide movement at the *joints,* where as the muscles attached to the bones produce their actions as they are stimulated to contract.

The *nervous system* is anatomically divided into the *central nervous system* (CNS), which includes the *brain* and *spinal cord*, and the *peripheral nervous system* (PNS), which includes the *cranial nerves*, arising from the brain, and the *spinal nerves*, arising from the spinal cord. The *autonomic nervous system* (ANS) is a functional division of the nervous system devoted to regulation of involuntary activities of the body. The brain and spinal cord are the centers for integration and coordination of information. *Nerves*, composed of *neurons*, convey nerve impulses to and from the brain. *Sensory organs*, such as the eyes and ears, respond to impulses in the environment and convey sensations to the CNS. The nervous system functions with the *endocrine system* in coordinating body activities.

The *cardiovascular system* consists of the *heart, vessels* (both blood and lymphatic vessels), *blood*, and the tissues that produce the *blood*. The four-chambered human heart is enclosed by a *pericardial sac* within the thoracic cavity. *Arteries* and *arterioles* transport blood away from the heart, *capillaries* permeate the tissues and are the functional units for product exchange with the cells, and *venules* and *veins* transport blood toward the heart. *Lymphatic vessels* return interstitial fluid back to the circulatory system after first passing it through *lymph nodes* for cleansing. Blood cells are produced in the bone marrow, and once old and worn, they are broken down in the liver.

The *respiratory system* consists of the *conducting division* that transports air to and from the *respiratory division* within the *lungs*. The *alveoli* of the lungs contact the capillaries of the cardiovascular system and are the sites for transport of respiratory gases into and out of the body.

The *digestive system* consists of a *gastrointestinal tract* (GI tract) and *accessory digestive organs*. Food traveling through the GI tract is processed such that it is suitable for absorption through the intestinal wall into the bloodstream. The *pancreas* and *liver* are the principal digestive organs. The pancreas produces hormones and enzymes. The liver processes nutrients, stores glucose as glycogen, and excretes bile.

Because of commonality of prenatal development and dual functions of some of the organs, the *urinary system* and *reproductive system* may be considered together as the *urogenital system*. The urinary system, consisting of the *kidneys, ureters, urinary bladder*, and *urethra*, extracts and processes wastes from the blood in the form of urine. The male and female reproductive systems produce regulatory hormones and gametes (sperm and ova, respectively) within the gonads (testes and ovaries). Sexual reproduction is the mechanism for propagation of offspring that have traits from both parents. The process of prenatal development is made possible by the formation of *extraembryonic membranes* (placenta, umbilical cord, allantois, amnion, chorion, and yolk sac) within the uterus of the pregnant woman.

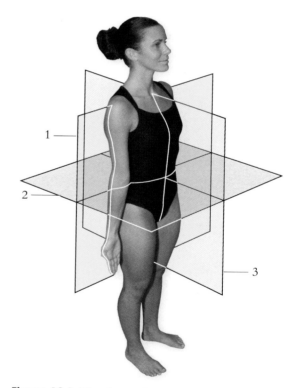

Figure 18.1 The planes of reference in a person while standing in anatomical position. The anatomical position provides a basis of reference for describing the relationship of one body part to another. In the anatomical position, the person is standing, the feet are parallel, the eyes are directed forward, and the arms are to the sides with the palms turned forward and the fingers pointed straight down.

1. Transverse plane (cross-sectional plane)
2. Coronal plane (frontal plane)
3. Sagittal plane

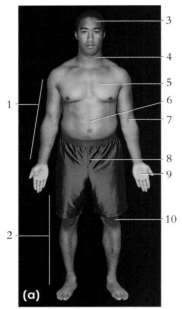
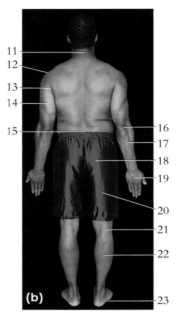

Figure 18.2 The major body parts and regions in humans (bipedal vertebrate). (a) An anterior view and (b) a posterior view.

1. Upper extremity
2. Lower extremity
3. Head
4. Neck, anterior aspect
5. Thorax (chest)
6. Abdomen
7. Cubital fossa
8. Pubic region
9. Palmar region (palm)
10. Patellar region (patella)
11. Cervical region
12. Deltoid region (shoulder)
13. Axilla (armpit)
14. Brachium (upper arm)
15. Lumbar region
16. Elbow
17. Antebrachium (forearm)
18. Gluteal region (buttock)
19. Dorsum of hand
20. Femoral region (thigh)
21. Popliteal fossa
22. Calf
23. Plantar surface (sole)

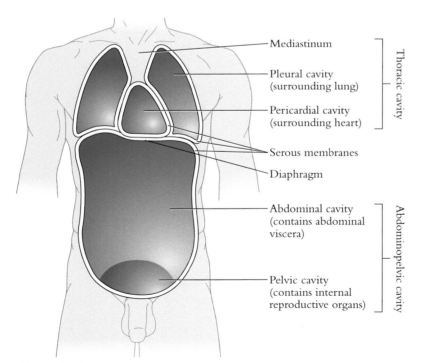

Figure 18.3 An anterior view (coronal plane) of the body cavities of the trunk.

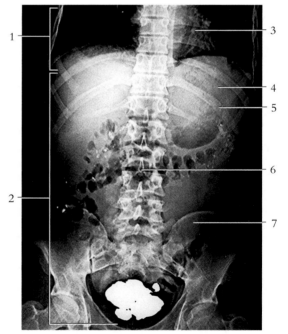

Figure 18.4 An MR image of the trunk showing the body cavities and their contents.

1. Thoracic cavity
2. Abdominopelvic cavity
3. Image of heart
4. Image of diaphragm
5. Image of rib
6. Image of lumbar vertebra
7. Image of ilium

Human Biology

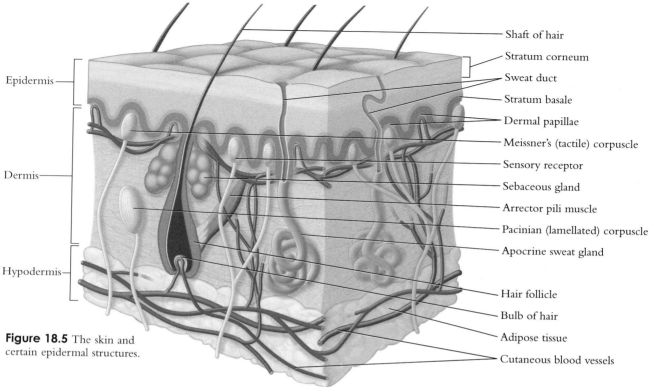

Figure 18.5 The skin and certain epidermal structures.

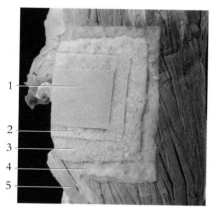

Figure 18.6 The gross structure of the skin and underlying fascia.
1. Epidermis
2. Dermis
3. Hypodermis
4. Fascia
5. Muscle

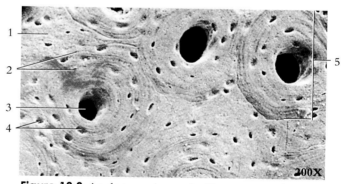

Figure 18.8 An electron micrograph of bone compact tissue.
1. Interstitial lamellae
2. Lamellae
3. Central canal (Haversian canal)
4. Lacunae
5. Osteon (Haversian system)

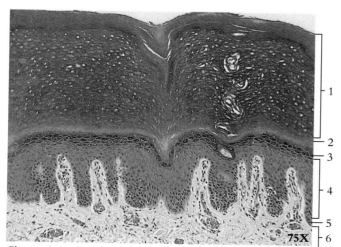

Figure 18.7 The epidermis and dermis of thick skin.
1. Stratum corneum
2. Stratum lucidum
3. Stratum granulosum
4. Stratum spinosum
5. Stratum basale
6. Dermis

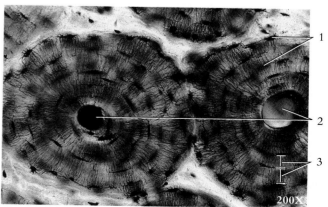

Figure 18.9 A transverse section of two osteons.
1. Lacunae with contained osteocytes
2. Central (Haversian) canals
3. Lamellae

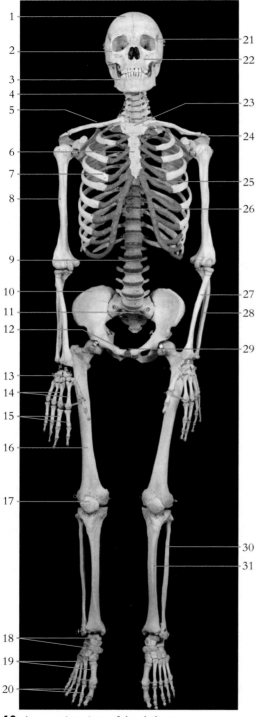

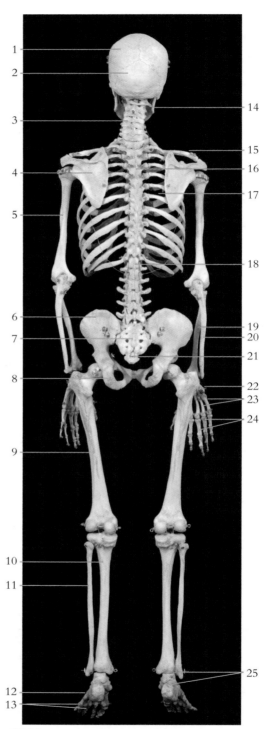

Figure 18.10 An anterior view of the skeleton.

1. Frontal bone
2. Zygomatic bone
3. Mandible
4. Cervical vertebra
5. Clavicle
6. Body of sternum
7. Rib
8. Humerus
9. Lumbar vertebra
10. Ilium
11. Sacrum
12. Pubis
13. Carpal bones
14. Metacarpal bones
15. Phalanges
16. Femur
17. Patella
18. Tarsal bones
19. Metatarsal bones
20. Phalanges
21. Orbit
22. Maxilla
23. Manubrium
24. Scapula
25. Costal cartilage
26. Thoracic vertebra
27. Radius
28. Ulna
29. Symphysis pubis
30. Fibula
31. Tibia

Figure 18.11 A posterior view of the skeleton.

1. Parietal bone
2. Occipital bone
3. Cervical vertebra
4. Scapula
5. Humerus
6. Ilium
7. Sacrum
8. Ischium
9. Femur
10. Tibia
11. Fibula
12. Metatarsal bones
13. Phalanges
14. Mandible
15. Clavicle
16. Thoracic vertebra
17. Rib
18. Lumbar vertebra
19. Radius
20. Ulna
21. Coccyx
22. Carpal bones
23. Metacarpal bones
24. Phalanges
25. Tarsal bones

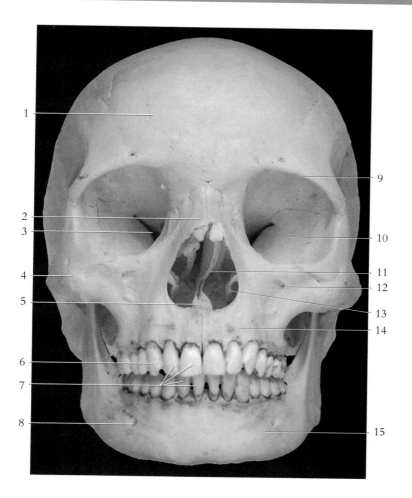

Figure 18.12 An anterior view of the skull.
1. Frontal bone
2. Nasal bone
3. Superior orbital fissure
4. Zygomatic bone
5. Vomer
6. Canine
7. Incisors
8. Mental foramen
9. Supraorbital margin
10. Sphenoid bone (orbital surface)
11. Perpendicular plate of ethmoid bone
12. Infraorbital foramen
13. Inferior nasal concha
14. Maxilla
15. Mandible

Figure 18.13 A lateral view of the skull.
1. Coronal suture
2. Frontal bone
3. Lacrimal bone
4. Nasal bone
5. Zygomatic bone
6. Maxilla
7. Premolars
8. Molars
9. Mandible
10. Parietal bone
11. Squamosal suture
12. Temporal bone
13. Lambdoidal suture
14. External acoustic meatus
15. Occipital bone
16. Condylar process of mandible
17. Mandibular notch
18. Mastoid process of temporal bone
19. Coronoid process of mandible
20. Angle of mandible

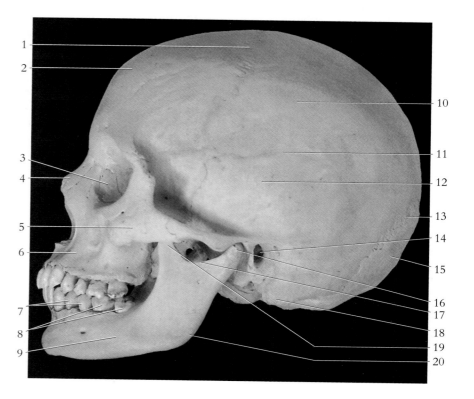

A Photographic Atlas for the Zoology Laboratory

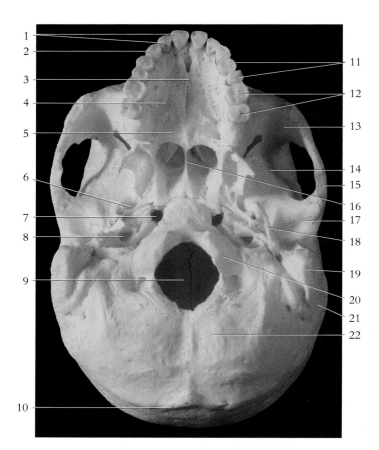

Figure 18.14 An inferior view of the skull.
1. Incisors
2. Canine
3. Intermaxillary suture
4. Maxilla
5. Palatine bone
6. Foramen ovale
7. Foramen lacerum
8. Carotid canal
9. Foramen magnum
10. Superior nuchal line
11. Premolars
12. Molars
13. Zygomatic bone
14. Sphenoid bone
15. Zygomatic arch
16. Vomer
17. Mandibular fossa
18. Styloid process of temporal bone
19. Mastoid process of temporal bone
20. Occipital condyle
21. Temporal bone
22. Occipital bone

Figure 18.15 A sagittal view of the skull.
1. Frontal bone
2. Frontal sinus
3. Crista galli of ethmoid bone
4. Cribriform plate of ethmoid bone
5. Nasal bone
6. Nasal concha
7. Maxilla
8. Mandible
9. Parietal bone
10. Occipital bone
11. Internal acoustic meatus
12. Sella turcica
13. Hypoglossal canal
14. Sphenoidal sinus
15. Styloid process of temporal bone
16. Vomer

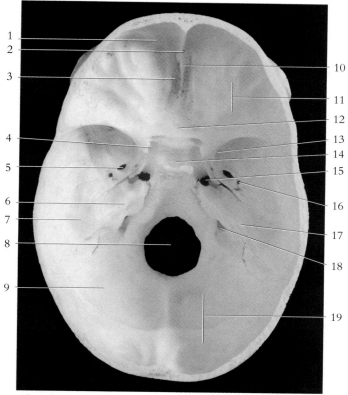

Figure 18.16 A superior view of the cranium.

1. Frontal bone
2. Foramen cecum
3. Cribriform plate of ethmoid bone
4. Optic canal
5. Foramen ovale
6. Petrous part of temporal bone
7. Temporal bone
8. Foramen magnum
9. Occipital bone
10. Crista galli of ethmoid bone
11. Anterior cranial fossa
12. Sphenoid bone
13. Foramen rotundum
14. Sella turcica of sphenoid bone
15. Foramen lacerum
16. Foramen spinosum
17. Internal acoustic meatus
18. Jugular foramen
19. Posterior cranial fossa

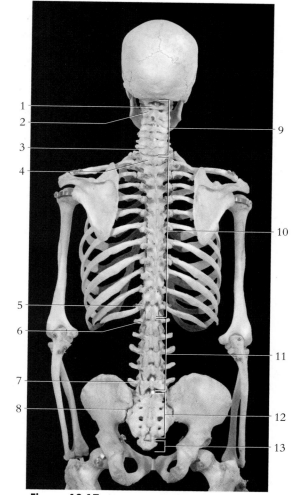

Figure 18.17 A posterior view of the vertebral column.

1. Atlas
2. Axis
3. 7th cervical vertebra
4. 1st thoracic vertebra
5. 12th thoracic vertebra
6. 1st lumbar vertebra
7. 5th lumbar vertebra
8. Sacroiliac joint
9. Cervical vertebrae
10. Thoracic vertebrae
11. Lumbar vertebrae
12. Sacrum
13. Coccyx

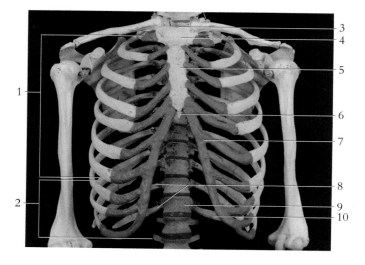

Figure 18.18 An anterior view of the rib cage.

1. True ribs (seven pairs)
2. False ribs (five pairs)
3. Jugular notch
4. Manubrium
5. Body of sternum
6. Xiphoid process
7. Costal cartilage
8. Floating ribs (inferior two pairs of false ribs)
9. 12th thoracic vertebra
10. 12th rib

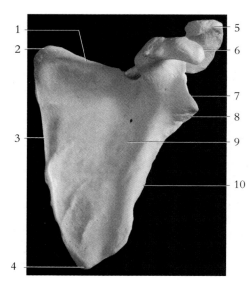

Figure 18.19 An anterior view of the left scapula.
1. Superior border
2. Superior angle
3. Medial (vertebral) border
4. Inferior angle
5. Acromion
6. Coracoid process
7. Glenoid fossa
8. Infraglenoid tubercle
9. Subscapular fossa
10. Lateral (axillary) border

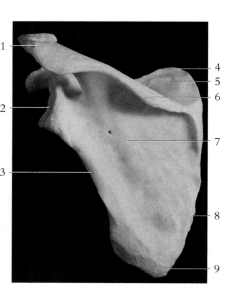

Figure 18.20 A posterior view of the left scapula.
1. Acromion
2. Glenoid fossa
3. Lateral (axillary) border
4. Superior angle
5. Supraspinous fossa
6. Spine
7. Infraspinous fossa
8. Medial (vertebral) border
9. Inferior angle

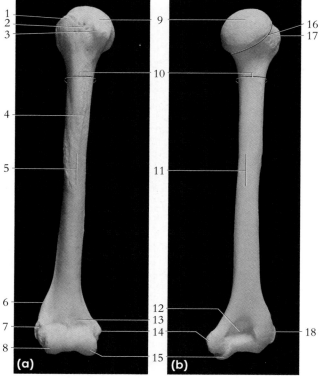

Figure 18.21 The right humerus. (a) Anterior view and (b) posterior view.
1. Greater tubercle
2. Intertubercular groove
3. Lesser tubercle
4. Deltoid tuberosity
5. Anterior body (shaft) of humerus
6. Lateral supracondylar ridge
7. Lateral epicondyle
8. Capitulum
9. Head of humerus
10. Surgical neck
11. Posterior body (shaft) of humerus
12. Olecranon fossa
13. Coronoid fossa
14. Medial epicondyle
15. Trochlea
16. Anatomical neck
17. Greater tubercle
18. Lateral epicondyle

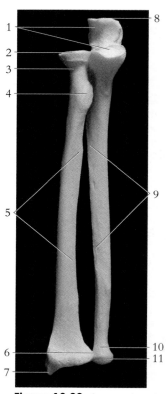

Figure 18.22 An anterior view of the right ulna and radius.
1. Trochlear notch
2. Head of radius
3. Neck of radius
4. Radial tuberosity
5. Interosseous margin
6. Location of ulnar notch of radius
7. Styloid process of radius
8. Olecranon
9. Interosseous margin
10. Neck of ulna
11. Head of ulna

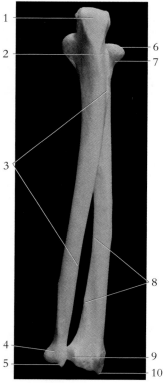

Figure 18.23 A posterior view of the right ulna and radius.
1. Olecranon
2. Location of radial notch of ulna
3. Interosseous margin
4. Head of ulna
5. Styloid process of ulna
6. Head of radius
7. Neck of radius
8. Interosseous margin
9. Ulnar notch
10. Styloid process of radius

Human Biology 205

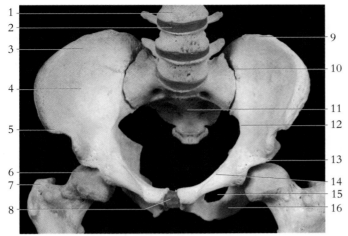

Figure 18.24 An anterior view of the articulated pelvic girdle showing the two coxal bones, the sacrum, and the two femora.

1. Lumbar vertebra
2. Intervertebral disk
3. Ilium
4. Iliac fossa
5. Anterior superior iliac spine
6. Head of femur
7. Greater trochanter
8. Symphysis pubis
9. Crest of the ilium
10. Sacroiliac joint
11. Sacrum
12. Pelvic brim
13. Acetabulum
14. Pubic crest
15. Obturator foramen
16. Ischium

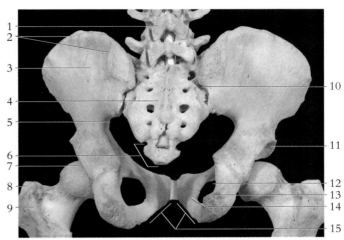

Figure 18.25 A posterior view of the articulated pelvic girdle showing the two coxal bones, the sacrum, and the two femora.

1. Lumbar vertebra
2. Crest of ilium
3. Ilium
4. Sacrum
5. Greater sciatic notch
6. Coccyx
7. Head of femur
8. Greater trochanter
9. Intertrochanteric crest
10. Sacroiliac joint
11. Acetabulum
12. Obturator foramen
13. Ischium
14. Pubis
15. Pubic angle

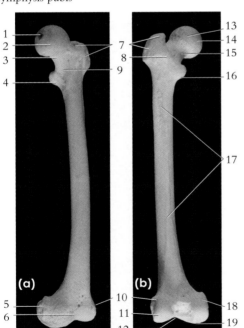

Figure 18.26 The left femur. (a) An anterior view and (b) a posterior view.

1. Fovea capitis femoris
2. Head
3. Neck
4. Lesser trochanter
5. Medial epicondyle
6. Patellar surface
7. Greater trochanter
8. Intertrochanteric crest
9. Intertrochanteric line
10. Lateral epicondyle
11. Lateral condyle
12. Intercondylar fossa
13. Head
14. Fovea capitis femoris
15. Neck
16. Lesser trochanter
17. Linea aspera on shaft (body) of femur
18. Medial epicondyle
19. Medial condyle

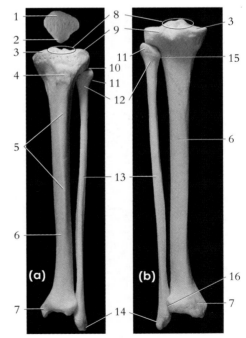

Figure 18.27 An anterior view of the (a) left patella, tibia, and fibula. (b) A posterior view of the left tibia and fibula.

1. Base of patella
2. Apex of patella
3. Medial condyle
4. Tibial tuberosity
5. Anterior crest of tibia
6. Body (shaft) of tibia
7. Medial malleolus
8. Intercondylar tubercles
9. Lateral condyle
10. Tibial articular facet of fibula
11. Head of fibula
12. Neck of fibula
13. Body (shaft) of fibula
14. Lateral malleolus
15. Fibular articular facet of tibia
16. Fibular notch of tibia

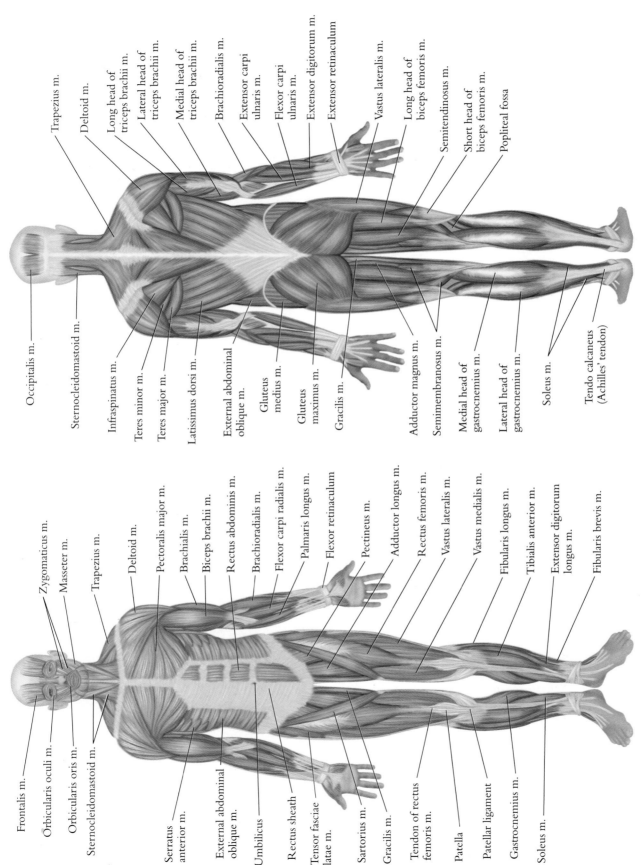

Figure 18.29 A posterior view of musculature (m = muscle).

Figure 18.28 An anterior view of musculature (m = muscle).

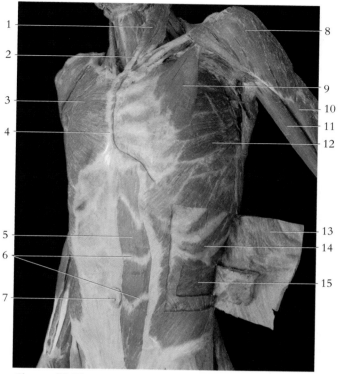

Figure 18.30 An anterolateral view of the trunk (m=muscle).
1. Sternocleidomastoid m.
2. Tendon of sternocleidomastoid m.
3. Pectoralis major m.
4. Sternum
5. Rectus abdominis m.
6. Tendinous inscriptions of rectus abdominis m.
7. Umbilicus
8. Deltoid m.
9. Pectoralis minor m.
10. Brachialis m.
11. Biceps brachii m. (long head)
12. Serratus anterior m.
13. External abdominal oblique m. (reflected)
14. External intercostal m.
15. Transverse abdominis m.

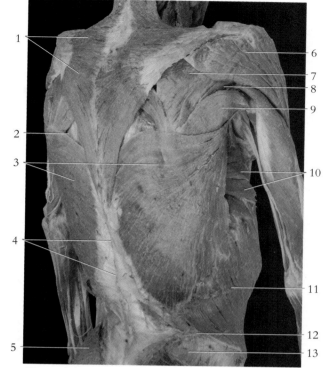

Figure 18.31 A posterolateral view of the trunk (m=muscle).
1. Trapezius m.
2. Triangle of auscultation
3. Latissimus dorsi mm.
4. Vertebral column (spinous processes)
5. Gluteus maximus m.
6. Deltoid m.
7. Infraspinatus m.
8. Teres minor m.
9. Teres major m.
10. Serratus anterior mm.
11. External abdominal oblique m.
12. Iliac crest
13. Gluteus medius m.

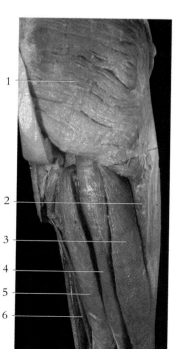

Figure 18.32 The superficial muscles of gluteal and femoral regions (m=muscle).
1. Gluteus maximus m.
2. Vastus lateralis m.
3. Biceps femoris m.
4. Semitendinosus m.
5. Semimembranosus m.
6. Gracilis m.

Figure 18.33 The deep structures of gluteal region (m=muscle).
1. Piriformis m.
2. Sciatic n.
3. Obturator internus m.
4. Quadratus femoris m.
5. Adductor minimus m.
6. Gluteus medius m. (reflected)
7. Gluteus minimus m.
8. Superior gemellus m.
9. Inferior gemellus m.
10. Gluteus maximus m. (reflected)

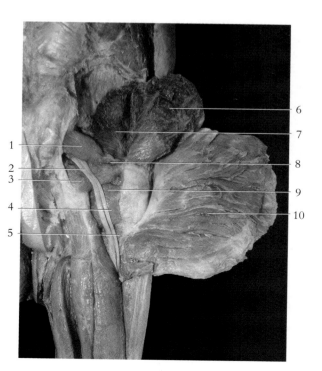

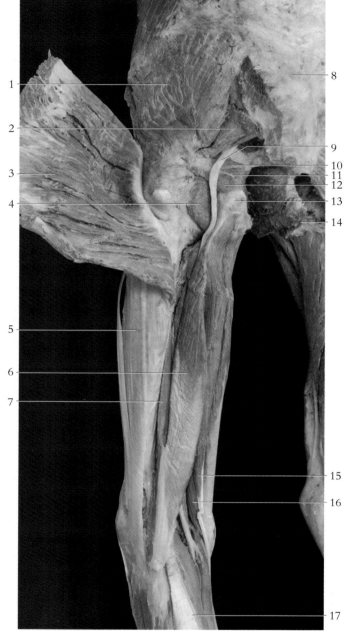

Figure 18.34 A posterolateral view of the posterior structures of femoral region (m=muscle).

1. Gluteus medius m.
2. Piriformis m.
3. Gluteus maximus m. (left side) (cut and reflected)
4. Quadratus femoris m.
5. Iliotibial band
6. Biceps femoris m. (long head)
7. Biceps femoris m. (short head).
8. Sacrum
9. Sciatic nerve
10. Superior gemellus m.
11. Obturator internus m.
12. Inferior gemellus m.
13. Ischial tuberosity
14. Anus
15. Semitendinosus m.
16. Semimembranosus m.
17. Gastrocnemius m.

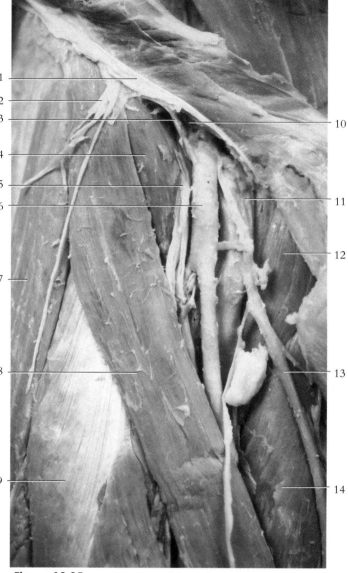

Figure 18.35 Anterior view of the right superior thigh (m=muscle).

1. Inguinal ligament
2. Lateral femoral cutaneous nerve
3. Superficial circumflex iliac artery
4. Iliopsoas m.
5. Femoral nerve
6. Femoral artery
7. Tensor fasciae latae m.
8. Sartorius m.
9. Rectus femoris m.
10. Femoral ring
11. Femoral vein
12. Pectineus m.
13. Great saphenous vein
14. Adductor longus m.

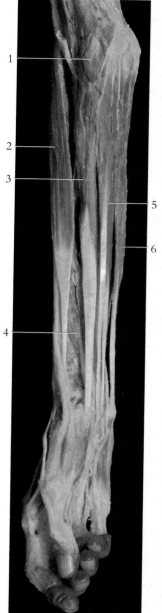

Figure 18.36 An anterior view of the superficial muscles of the right forearm (m=muscle).
1. Pronator teres m.
2. Brachioradialis m.
3. Flexor carpi radialis m.
4. Flexor pollicis longus m.
5. Palmaris longus m.
6. Flexor carpi ulnaris m.

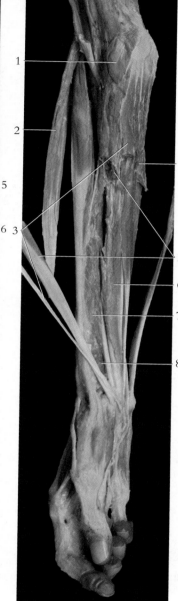

Figure 18.37 An anterior view of the deep muscles of the right forearm (m=muscle).
1. Pronator teres m.
2. Brachioradialis m. (cut and reflected)
3. Palmaris longus m. (cut and reflected)
4. Flexor carpi ulnaris m. (cut)
5. Flexor carpi radialis m. (cut and reflected)
6. Flexor digitorum superficialis m.
7. Flexor pollicis longus m.
8. Pronator quadratus m.

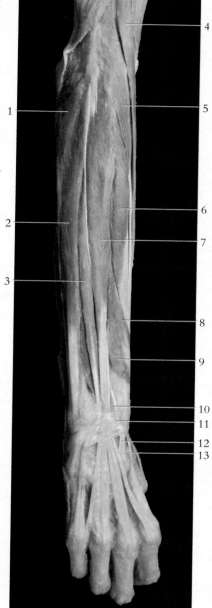

Figure 18.38 A posterior view of the superficial muscles of the right forearm (m=muscle).
1. Anconeus m.
2. Extensor carpi ulnaris m.
3. Extensor digiti minimi m.
4. Brachioradialis m.
5. Extensor carpi radialis longus m.
6. Extensor carpi radialis brevis m.
7. Extensor digitorum m.
8. Abductor pollicis longus m.
9. Extensor pollicis brevis m.
10. Extensor pollicis longus m.
11. Extensor retinaculum
12. Tendon of extensor carpi radialis brevis
13. Tendon of extensor carpi radialis longus

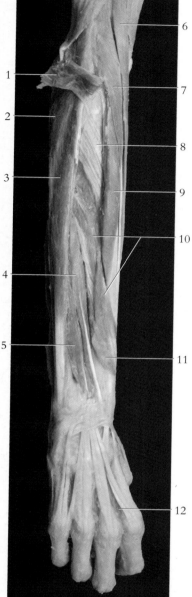

Figure 18.39 A posterior view of the deep muscles of the right forearm (m=muscle).
1. Extensor digitorum m. (cut and reflected)
2. Anconeus m.
3. Extensor carpi ulnaris m.
4. Extensor pollicis longus m.
5. Extensor indicis m.
6. Brachioradialis m.
7. Extensor carpi radialis longus m.
8. Supinator m.
9. Extensor carpi radialis brevis m.
10. Abductor pollicis longus m.
11. Extensor pollicis brevis m.
12. Dorsal interosseous m.

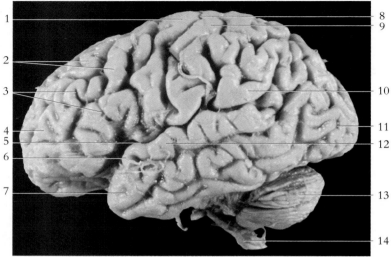

Figure 18.40 A lateral view of the brain.

1. Primary motor cerebral cortex
2. Gyri
3. Sulci
4. Frontal lobe of cerebrum
5. Lateral sulcus
6. Olfactory cerebral cortex
7. Temporal lobe of cerebrum
8. Central sulcus
9. Primary sensory cerebral cortex
10. Parietal lobe of cerebrum
11. Occipital lobe of cerebrum
12. Auditory cerebral cortex
13. Cerebellum
14. Medulla oblongata

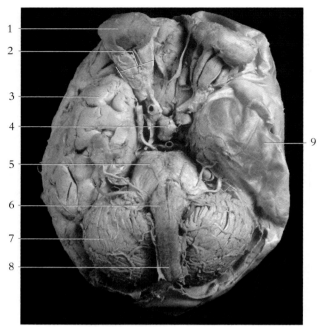

Figure 18.41 An inferior view of the brain with the eyes and part of the meninges still intact.

1. Eyeball
2. Muscles of the eye
3. Temporal lobe of cerebrum
4. Pituitary gland
5. Pons
6. Medulla oblongata
7. Cerebellum
8. Spinal cord
9. Dura mater

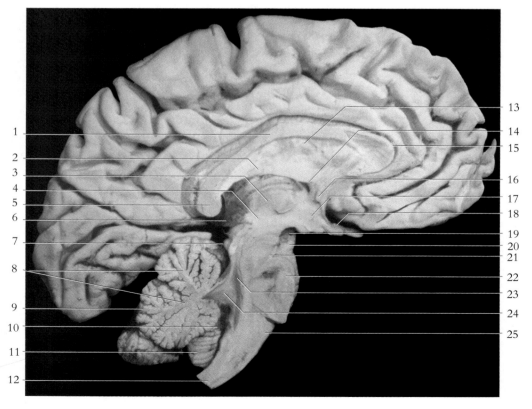

Figure 18.42 A sagittal view of the brain.

1. Body of corpus callosum
2. Crus of fornix
3. 3rd ventricle
4. Posterior commissure
5. Splenium of corpus callosum
6. Pineal body
7. Inferior colliculus
8. Arbor vitae of cerebellum
9. Vermis of cerebellum
10. Choroid plexus of 4th ventricle
11. Tonsilla of cerebellum
12. Medulla oblongata
13. Septum pellucidum
14. Intraventricular foramen
15. Genu of corpus callosum
16. Anterior commissure
17. Hypothalamus
18. Optic chiasma
19. Oculomotor nerve
20. Cerebral peduncle
21. Midbrain
22. Pons
23. Mesencephalic (cerebral) aqueduct
24. 4th ventricle
25. Pyramid of medulla oblongata

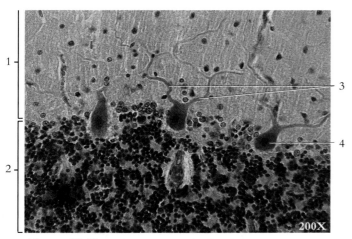

Figure 18.43 A photomicrograph of Purkinje neurons from the cerebellum.
1. Molecular layer of cerebellar cortex
2. Granular layer of cerebellar cortex
3. Dendrites of Purkinje cell
4. Purkinje cell body

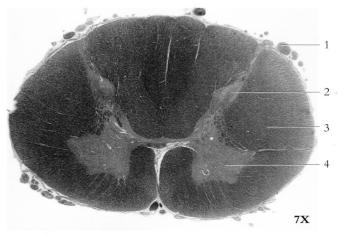

Figure 18.44 A transverse section of the spinal cord.
1. Posterior (dorsal) root of spinal nerve
2. Posterior (dorsal) horn (gray matter)
3. Spinal cord tract (white matter)
4. Anterior (ventral) horn (gray matter)

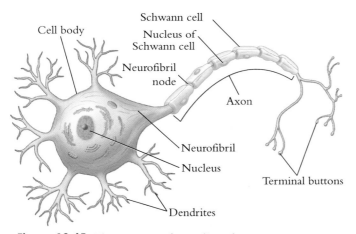

Figure 18.45 The structure of a myelinated neuron.

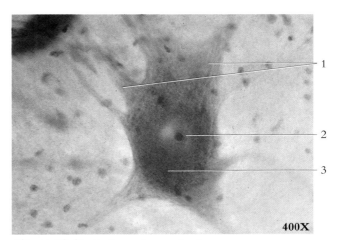

Figure 18.46 A photomicrograph of a neuron.
1. Cytoplasmic extensions
2. Nucleolus
3. Cell body of neuron

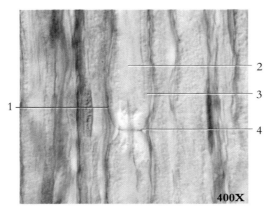

Figure 18.47 The histology of a myelinated nerve.
1. Endoneurium
2. Axon
3. Myelin layer
4. Neurofibril node (node of Ranvier)

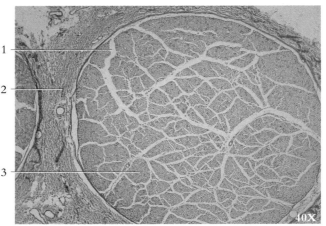

Figure 18.48 A transverse section of a nerve.
1. Perineurium
2. Epineurium
3. Bundle of axons (fascicle)

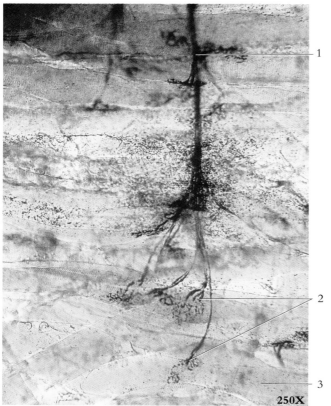

Figure 18.49 The neuromuscular junction.
1. Motor axon
2. Motor end plates
3. Skeletal muscle fiber

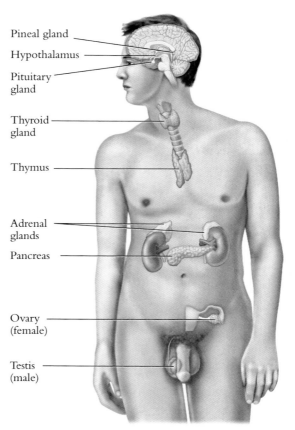

Figure 18.50 The principal endocrine glands.

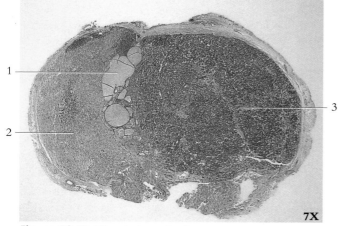

Figure 18.51 The pituitary gland.
1. Pars intermedia (adenohypophysis)
2. Pars nervosa (neurohypophysis)
3. Pars distalis (adenohypophysis)

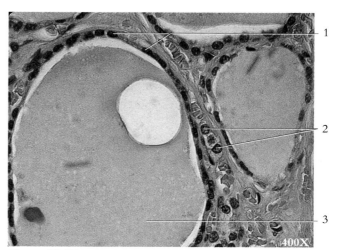

Figure 18.52 The thyroid gland.
1. Follicle cells
2. C cells
3. Colloid within follicle

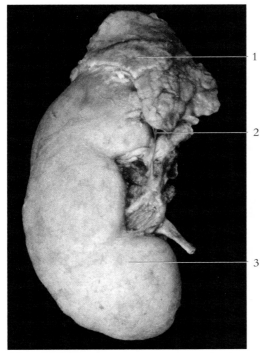

Figure 18.53 The adrenal (suprarenal) gland.
1. Adrenal gland
2. Inferior suprarenal artery
3. Kidney

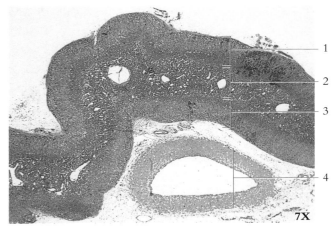

Figure 18.54 The adrenal gland.
1. Adrenal cortex
2. Adrenal medulla
3. Adrenal cortex
4. Blood vessel

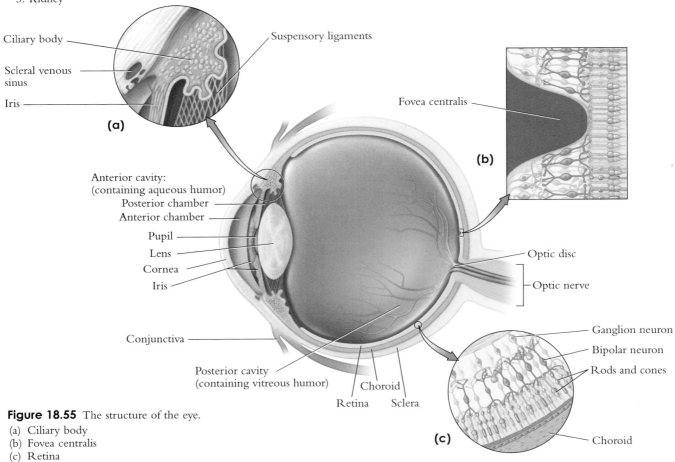

Figure 18.55 The structure of the eye.
(a) Ciliary body
(b) Fovea centralis
(c) Retina

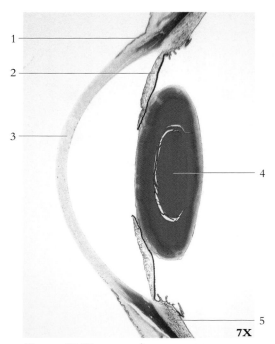

Figure 18.56 A sagittal section of the anterior portion of the eye.
1. Conjunctiva
2. Iris
3. Cornea
4. Lens
5. Ciliary body

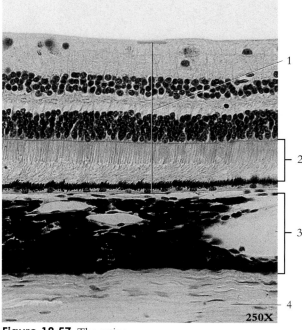

Figure 18.57 The retina.
1. Retina
2. Rods and cones
3. Choroid
4. Sclera

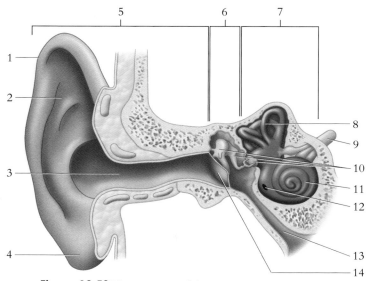

Figure 18.58 The structure of the ear.
1. Helix
2. Auricle
3. External auditory canal
4. Earlobe
5. Outer ear
6. Middle ear
7. Inner ear
8. Semicircular canals
9. Vestibulocochlear nerve
10. Auditory ossicles
11. Cochlea
12. Vestibular (oval) window
13. Auditory tube
14. Tympanic membrane

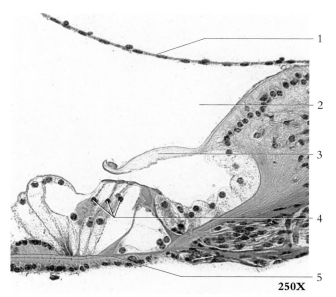

Figure 18.59 The spiral organ (organ of Corti).
1. Vestibular membrane
2. Cochlear duct
3. Tectorial membrane
4. Hair cells
5. Basilar membrane

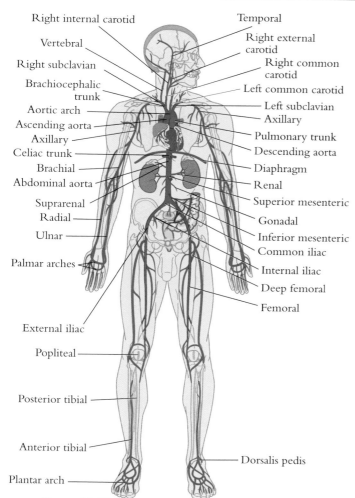

Figure 18.60 The principal arteries of the body.

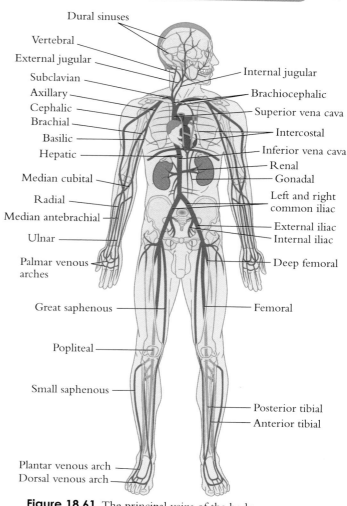

Figure 18.61 The principal veins of the body.

Figure 18.62 The position of the heart within the pericardium.
1. Mediastinum
2. Right lung
3. Pericardium
4. Diaphragm
5. Liver
6. Left lung

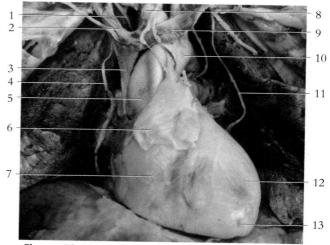

Figure 18.63 Anterior view of the heart and associated structures.
1. Right vagus nerve
2. Right brachiocephalic vein
3. Superior vena cava
4. Right phrenic nerve
5. Ascending aorta
6. Pericardium (cut)
7. Right ventricle of heart
8. Brachiocephalic artery
9. Left brachiocephalic vein
10. Aortic arch
11. Left phrenic nerve
12. Left ventricle of heart
13. Apex of heart

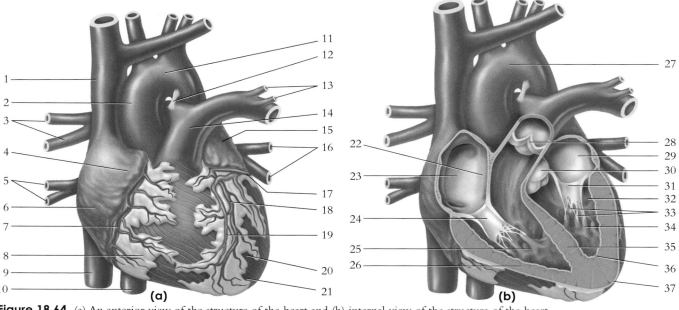

Figure 18.64 (a) An anterior view of the structure of the heart and (b) internal view of the structure of the heart.

1. Superior vena cava
2. Ascending aorta
3. Branches of right pulmonary artery
4. Auricle of right atrium
5. Right pulmonary veins
6. Right atrium
7. Right coronary artery and vein
8. Right ventricle
9. Inferior vena cava
10. Thoracic aorta
11. Aortic arch
12. Ligamentum arteriosum
13. Branches of left pulmonary artery
14. Pulmonary trunk
15. Left atrium
16. Left pulmonary veins
17. Circumflex artery
18. Anterior interventricular artery
19. Anterior interventricular vein
20. Left ventricle
21. Apex of heart
22. Interatrial septum
23. Right atrium
24. Tricuspid valve
25. Right ventricle
26. Myocardium
27. Aortic arch
28. Pulmonary valve
29. Left atrium
30. Aortic valve
31. Bicuspid valve
32. Left ventricle
33. Chordae tendinae
34. Papillary muscle
35. Interventricular septum
36. Endocardium
37. Visceral pericardium

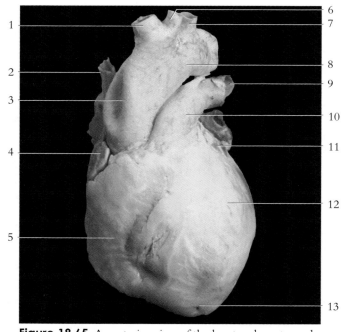

Figure 18.65 An anterior view of the heart and great vessels.

1. Brachiocephalic trunk
2. Superior vena cava
3. Ascending aorta
4. Right atrium
5. Right ventricle
6. Left common carotid artery
7. Left subclavian artery
8. Aortic arch
9. Pulmonary artery
10. Pulmonary trunk
11. Left atrium
12. Left ventricle
13. Apex of heart

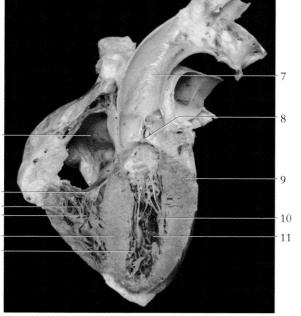

Figure 18.66 The internal structure of the heart.

1. Right atrium
2. Right atrioventricular valve
3. Chordae tendinae
4. Right ventricle
5. Interventricular septum
6. Trabeculae carneae
7. Ascending aorta
8. Aortic valve
9. Myocardium
10. Papillary muscle
11. Left ventricle

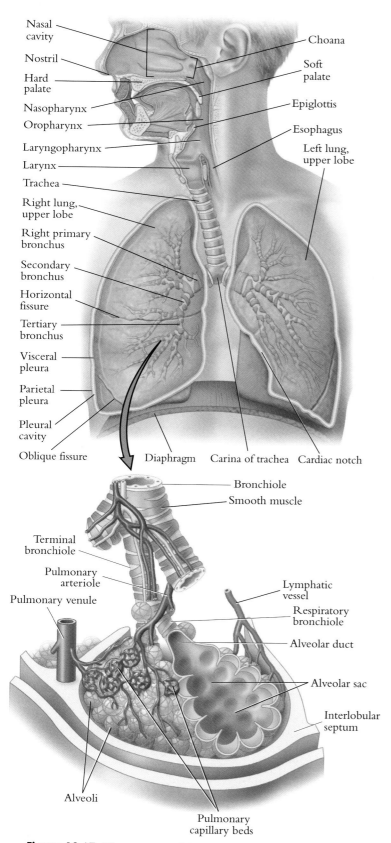

Figure 18.67 The structure of the respiratory system.

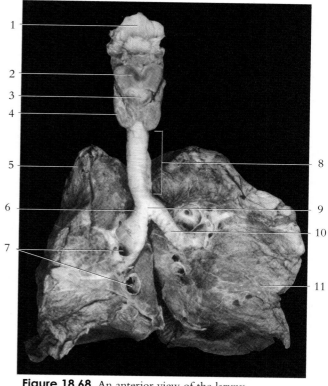

Figure 18.68 An anterior view of the larynx, trachea, and lungs.

1. Epiglottis
2. Thyroid cartilage
3. Cricoid cartilage
4. Thyroid gland
5. Right lung
6. Right principal (primary) bronchus
7. Pulmonary vessels
8. Trachea
9. Carina
10. Left principal (primary) bronchus
11. Left lung

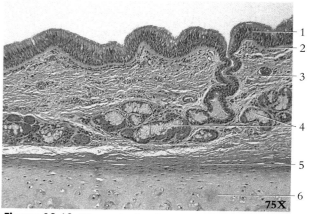

Figure 18.69 The tracheal wall.

1. Respiratory epithelium
2. Basement membrane
3. Duct of seromucous gland
4. Seromucous glands
5. Perichondrium
6. Hyaline cartilage

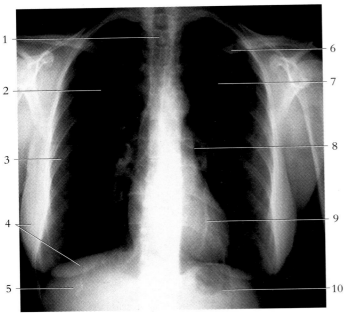

Figure 18.70 A radiograph of the thorax.
1. Thoracic vertebra
2. Right lung
3. Rib
4. Image of right breast
5. Diaphragm/liver
6. Clavicle
7. Left lung
8. Mediastinum
9. Heart
10. Diaphragm/stomach

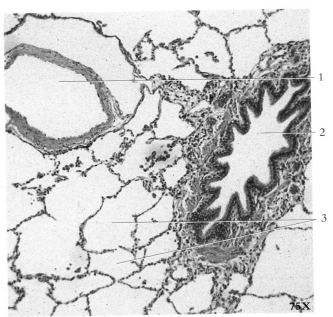

Figure 18.71 A bronchiole.
1. Pulmonary arteriole
2. Bronchiole
3. Pulmonary alveoli

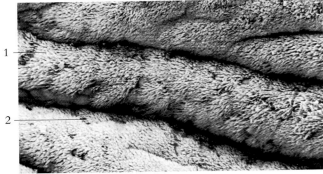

Figure 18.72 An electron micrograph of the lining of the trachea.
1. Cilia
2. Goblet cell

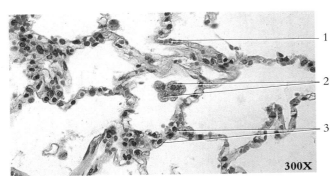

Figure 18.73 The pulmonary alveoli.
1. Capillary in alveolar wall
2. Macrophages
3. Type II pneumocytes

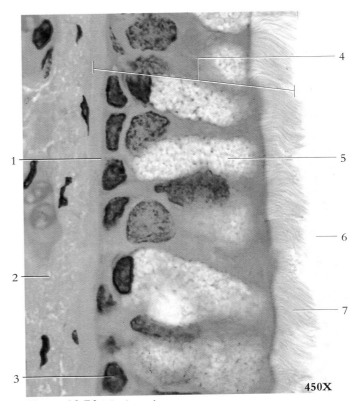

Figure 18.74 The bronchus.
1. Basement membrane
2. Lamina propria
3. Nucleus of epithelial cell
4. Pseudostratified columnar epithelium
5. Goblet cell
6. Lumen of bronchus
7. Cilia

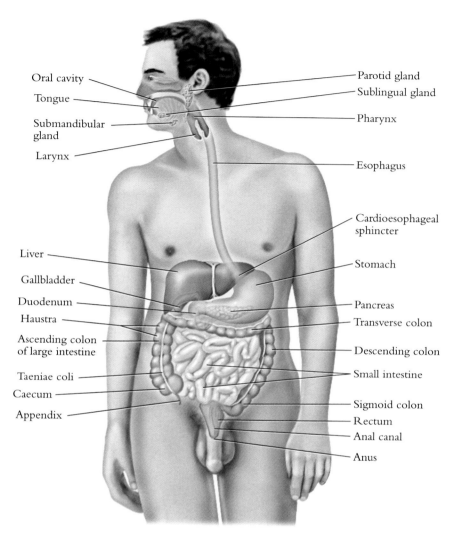

Figure 18.75 The structure of the digestive system.

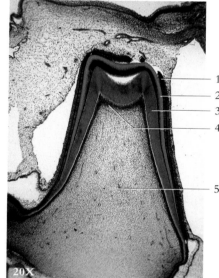

Figure 18.76 A developing tooth.
1. Ameloblasts
2. Enamel
3. Dentin
4. Odontoblasts
5. Pulp

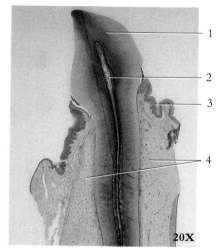

Figure 18.77 A mature tooth.
1. Dentin (enamel has been dissolved away)
2. Pulp
3. Gingiva
4. Alveolar bone

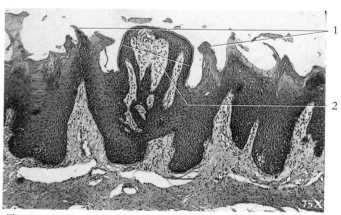

Figure 18.78 The filiform and fungiform papillae.
1. Filiform papillae
2. Fungiform papilla

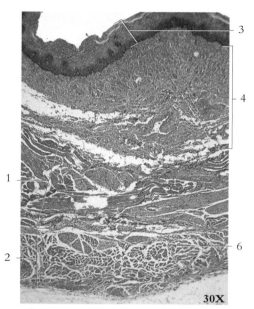

Figure 18.79 The wall of the esophagus.
1. Inner circular layer (muscularis externa)
2. Outer longitudinal layer (muscularis externa)
3. Mucosa
4. Submucosa
5. Muscularis externa

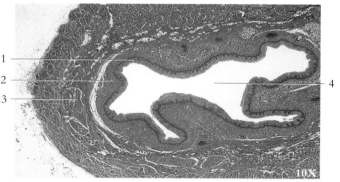

Figure 18.80 A transverse section of esophagus.
1. Mucosa
2. Submucosa
3. Muscularis
4. Lumen

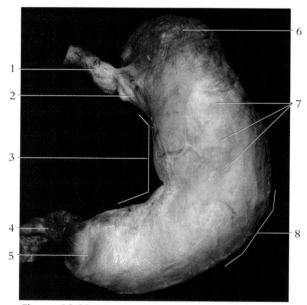

Figure 18.82 The major regions and structures of the stomach.
1. Esophagus
2. Cardiac portion of stomach
3. Lesser curvature of stomach
4. Duodenum
5. Pylorus of stomach
6. Fundus of stomach
7. Body of stomach
8. Greater curvature of stomach

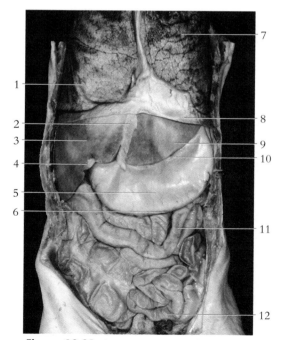

Figure 18.81 An anterior view of the trunk.
1. Right lung
2. Falciform ligament
3. Right lobe of liver
4. Gallbladder
5. Body of stomach
6. Greater curvature of stomach
7. Left lung
8. Diaphragm
9. Left lobe of liver
10. Lesser curvature of stomach
11. Transverse colon
12. Small intestine

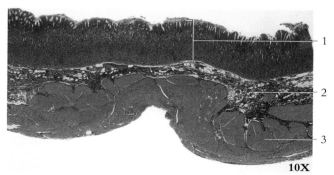

Figure 18.83 The wall of stomach.
1. Mucosa
2. Submucosa
3. Muscularis externa

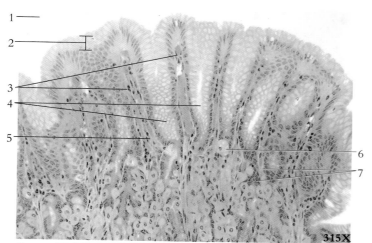

Figure 18.84 The histology of the cardiac region of the stomach.
1. Lumen of stomach
2. Surface epithelium
3. Mucosal ridges
4. Gastric pits
5. Lamina propria
6. Parietal cells
7. Chief (zymogenic) cells

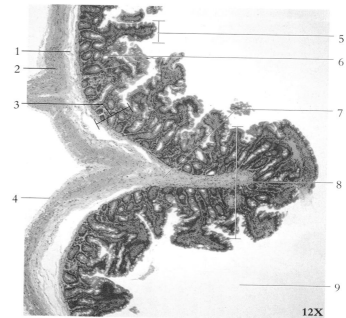

Figure 18.85 The histology of the jejunum of the small intestine.
1. Submucosa
2. Circular and longitudinal muscles
3. Mucosa
4. Serosa
5. Villus
6. Intestinal glands
7. Submucosa
8. Plica circulares
9. Lumen of small intestine

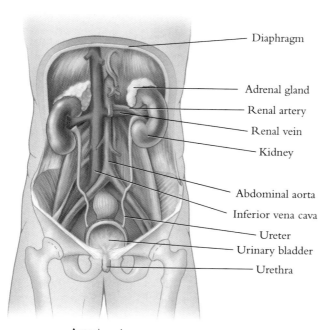

Anterior view

Figure 18.86 The organs of the urinary system.

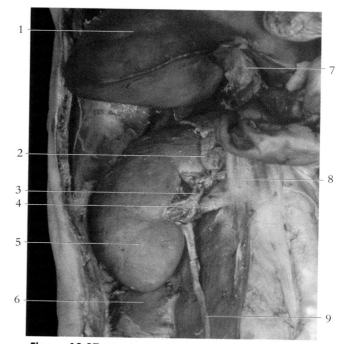

Figure 18.87 The kidney and ureter with overlying viscera removed.
1. Liver
2. Adrenal gland
3. Renal artery
4. Renal vein
5. Right kidney
6. Quadratus lumborum muscle
7. Gallbladder
8. Inferior vena cava
9. Ureter

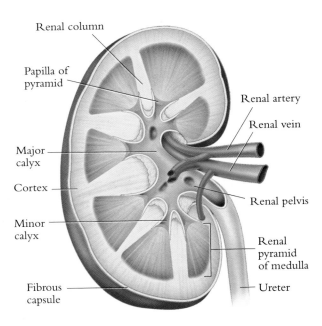

Figure 18.88 The structure of the kidney.

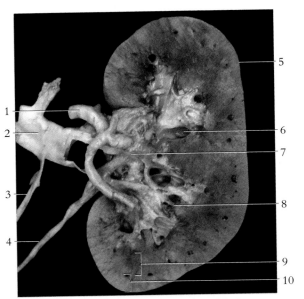

Figure 18.89 A coronal section of the left kidney.
1. Renal artery
2. Renal vein
3. Left testicular vein
4. Ureter
5. Renal capsule
6. Major calyx
7. Renal pelvis
8. Renal papilla
9. Renal medulla
10. Renal cortex

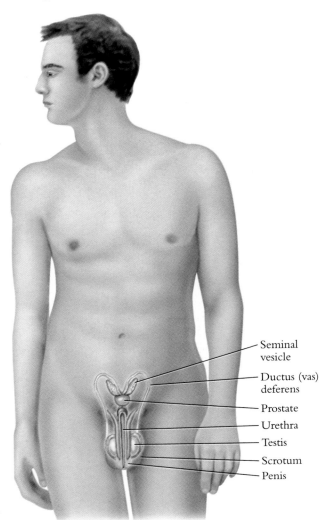

Figure 18.90 The organs of the male reproductive system.

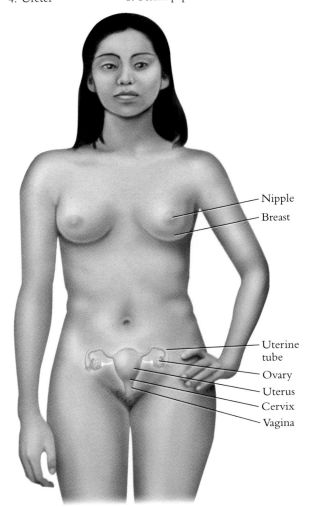

Figure 18.91 The organs of the female reproductive system.

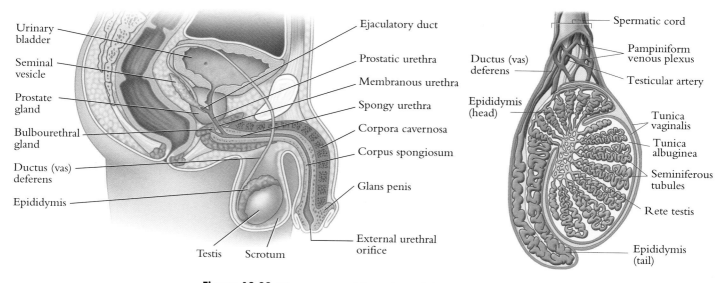

Figure 18.92 The structure of the male genitalia.

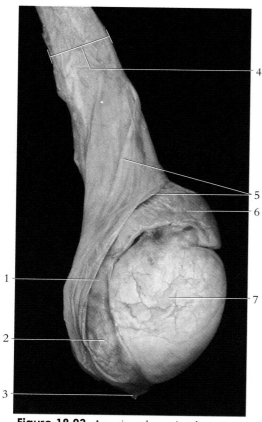

Figure 18.93 A testis and associated structures.
1. Body of epididymis
2. Tail of epididymis
3. Gubernaculum
4. Spermatic cord
5. Spermatic fascia
6. Head of epididymis
7. Testis

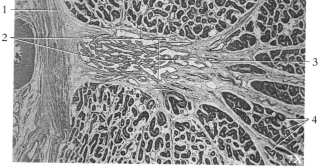

Figure 18.94 The testis.
1. Tunic albuginea
2. Tubules of rete testis
3. Mediastinum
4. Seminiferous tubules

Figure 18.95 The structure of the ovary.
1. Corona radiata
2. Secondary oocyte
3. Ovulation
4. Follicular fluid within antrum
5. Cumulus oophorus
6. Oocyte
7. Follicular cells
8. Germinal epithelium
9. Primary follicles
10. Ovarian vessels
11. Corpus albicans
12. Ovarian cortex
13. Ovarian medulla
14. Corpus luteum

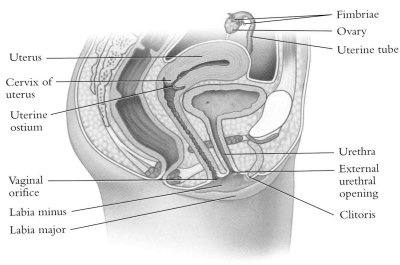

Figure 18.96 The external genitalia and internal reproductive organs of the female reproductive system.

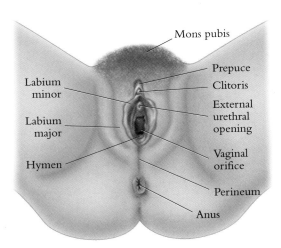

Figure 18.97 The female external genitalia (vulva).

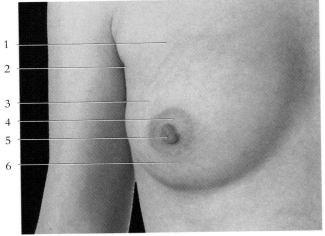

Figure 18.98 The surface anatomy of the female breast.
1. Pectoralis major muscle
2. Axilla
3. Lateral process of breast
4. Areola
5. Nipple
6. Breast (containing mammary glands)

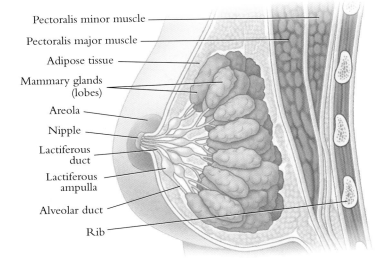

Figure 18.99 Internal structure of the female breast.

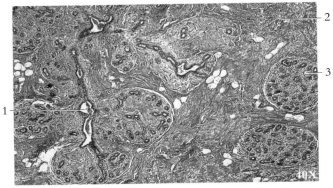

Figure 18.100 Mammary glands (nonlactating glands).
1. Interlobular duct
2. Interlobular connective tissue
3. Lobule of glandular tissue

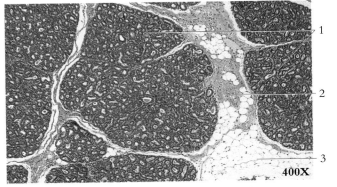

Figure 18.101 Mammary glands (lactating glands).
1. Lobules of glandular tissue
2. Interlobular connective tissue
3. Adipose cells

Glossary of Terms

A

abdomen: the region of the mammalian body located between the diaphragm and the pelvis, which contains the abdominal cavity and its visceral organs; one of the three principal body regions (head, thorax, and abdomen) of many animals.

abduction: a movement away from the axis or midline of the body; opposite of adduction.

abiotic: without living organisms; nonliving portions of the environment.

acapnia: a decrease in normal amount of CO_2 in the blood.

accommodation: a change in the shape of the lens of the eye so that vision is more acute; the focusing for various distances.

acetone: an organic compound that may be present in the urine of diabetics; also called ketone bodies.

acetylcholine: a neurotransmitter chemical secreted at the terminal ends of many axons, responsible for postsynaptic transmission; also called ACh.

acetylcholinesterase: an enzyme that breaks down acetylcholine; also called AChE.

Achilles' tendon: see *tendo calcaneus*.

acid: a substance that releases hydrogen ions (H^+) in a solution; a solution in which the pH is less than 7; acidic.

acidosis: a disorder of body chemistry in which the alkaline substances of the blood are reduced below normal.

acoelomate: without a coelomic cavity, as in flatworms.

acoustic: referring to sound or the sense of hearing.

actin: a protein in muscle fibers that together with myosin is responsible for contraction.

action potential: the change in ionic charge propagated along the membrane of a neuron; the nerve impulse.

active transport: movement of a substance into or out of a cell from a lesser to a greater concentration, requiring a carrier molecule and expenditure of energy.

adaptation: structural, physiological, or behavioral traits of an organism that promote its survival and contribute to its ability to reproduce under environmental conditions.

adduction: a movement toward the axis or midline of the body; opposite of abduction.

adenohypophysis: anterior pituitary gland.

adenoid: paired lymphoid structures in the nasopharynx; also called pharyngeal tonsils.

adenosine triphosphate (ATP): a chemical compound that provides energy for cellular use.

adipose: fat, or fat-containing, such as adipose tissue.

adrenal glands: endocrine glands; one superior to each kidney; also called suprarenal glands.

aerobic: requiring free O_2 for growth and metabolism as in the case of certain bacteria called aerobes.

agglutination: clumping of cells; particular reference to red blood cells in an antigen antibody reaction.

aggression: provoking, domineering behavior.

allantois: an extraembryonic membranous sac that forms blood cells and gives rise to the fetal umbilical arteries and vein. It also contributes to the formation of the urinary bladder.

allele: an alternative form of a gene occurring at a given chromosome site, or locus.

all-or-none response: functioning completely when exposed to a stimulus of threshold strength; applies to action potentials through neurons and muscle fiber contraction.

alpha helix: right-handed spiral typical in proteins and DNA.

altruism: behavior benefiting other organisms without regard to its possible advantage or detrimental effect on the performer.

alveolus: A capsule within a structure. Pulmonary alveoli are the basic functional units of respiration.

amino acid: a unit of protein that contains an amino group (NH_2) and an acid group (COOH).

amnion: an extraembryonic membrane that surrounds the fetus to contain the amniotic fluid.

amniote: an animal that has an amnion during embryonic development; reptiles, birds, and mammals.

amoeba: protozoans that move by means of pseudopodia.

amphiarthrosis: a slightly movable joint.

anaerobic respiration: metabolizing and growing in the absence of oxygen.

analogous: similar in function regardless of developmental origin; generally in reference to similar adaptations.

anatomical position: the position in human anatomy in which there is an erect body stance with the eyes directed forward, the arms at the sides, and the palms of the hands facing forward.

anatomy: the branch of science concerned with the structure of the body and the relationship of its organs.

annulus: a ringlike segment, such as body rings on leeches.

antebrachium: the forearm.

antenna: a sensory appendage on many species of invertebrate animals.

anterior (ventral): toward the front; the opposite of posterior (dorsal).

anticodon: three ("a triplet") nucleotides in transfer RNA that pair with a complementary codon (triplet) in messenger RNA.

antigen: a foreign material, usually a protein, that triggers the immune system to produce antibodies.

anus: the terminal end of the GI tract, opening of the anal canal.

aorta: the major systemic vessel of the arterial portion of the circulatory system, emerging from the left ventricle.

apocrine gland: a sweat gland that functions in body cooling; secretes pheromones.
apopyle: opening of the radial canal into the spongocoel of sponges.
appeasement: submissive behavior, usually soliciting an end to aggression.
appendix: a short pouch that attaches to the caecum.
aqueous humor: the watery fluid that fills the anterior and posterior chambers of the eye.
arbor vitae: the branching arrangement of white matter within the cerebellum.
archaebacteria: prokaryotic organisms that represent an early group of simple life forms, similar to bacteria but more closely related to eukaryotes.
archenteron: a principal cavity of an embryo during the gastrula stage. Lined with endoderm, the archenteron develops into the digestive tract.
areola: the pigmented ring around the nipple.
artery: a blood vessel that carries blood away from the heart.
articular cartilage: a hyaline cartilaginous covering over the articulating surface of bones of synovial joints.
ascending colon: the portion of the large intestine between the caecum and the hepatic flexure.
asexual: lacking distinct sexual organs and lacking the ability to produce gametes.
aster: minute rays of microtubules at the ends of the spindle apparatus in animal cells during cell division.
asymmetrical: not symmetrical.
atom: the smallest unit of an element that can exist and still have the properties of the element; collectively, atoms form molecules in a compound.
atomic number: the number of protons within the nucleus of an atom.
atomic weight: the number of protons together with the number of neutrons within the nucleus of an atom.
ATP: a compound of adenine, ribose, and three phosphates, two of which are high-energy phospates; it is the energy source for most cellular processes.
atrium: either of the two superior chambers of the heart that receive venous blood.
atrophy: wasting away or decrease in size of a cell or organ.
auditory tube: a narrow canal that connects the middle-ear chamber to the pharynx; also called the eustachian canal.
autonomic: self-governing; pertaining to the division of the nervous system that controls involuntary activities.
autosome: a chromosome other than a sex chromosome.
autotroph: an organism capable of synthesizing its own organic molecules (food) from inorganic molecules.
axilla: the depressed hollow under the arm; the armpit.
axon: The elongated process of a neuron (nerve cell) that transmits an impulse away from the cell body.

B

bacteria: prokaryotes within the domain Bacteria, lacking the organelles of eukaryotic cells.
base: a substance that contributes or liberates hydroxide ions in a solution; a solution in which the pH is greater than 7; alkaline.
basement membrane: a thin sheet of extracellular substance to which the basal surfaces of epithelial cells are attached.
basophil: a granular leukocyte that readily stains with basophilic dye.
belly: the thickest circumference of a skeletal muscle.
benign: nonmalignant; a confined tumor.
bilateral symmetry: the morphologic condition of having similar right and left halves.
binary fission: a process of reproduction that does not involve a mitotic spindle.
binomial system: assignment of two names to an organism, the first of which is the genus and the second the specific epithet, together constituting the species name.
biome: a major climax community characterized by a particular group of plants and animals.
biosphere: the portion of the Earth's atmosphere and surface where living organisms exist.
biotic: pertaining to aspects of life, especially to characteristics of populations or ecosystems.
blastocoel: the cavity of a blastocyst.
blastula: an early stage of prenatal development between the morula and embryonic stages.
blood: the fluid connective tissue that circulates through the cardiovascular system to transport substances throughout the body.
bolus: a moistened mass of food that is swallowed from the oral cavity into the pharynx.
bone: an organ composed of solid, rigid connective tissue, forming a component of the skeletal system.
Bowman's capsule: see *glomerular capsule*.
brain: the enlarged superior portion of the central nervous system, located in the cranial cavity of the skull.
brainstem: the portion of the brain consisting of the medulla oblongata, pons, and midbrain.
bronchial tree: the bronchi and their branching bronchioles.
bronchiole: a small division of a bronchus within the lung.
bronchus: a branch of the trachea that leads to a lung.
buccal cavity: the mouth, or oral cavity.
buffer: a compound or substance that prevents large changes in the pH of a solution.
bursa: a saclike structure filled with synovial fluid that occurs around joints.
buttock: the rump or fleshy mass on the posterior aspect of the lower trunk, formed primarily by the gluteal muscles.

C

caecum: the pouchlike portion of the large intestine to which the ileum of the small intestine is attached; also spelled cecum.
calorie: the heat required to raise one kilogram of water one degree centigrade.
calyx: a cup-shaped portion of the renal pelvis that encircles renal papillae.

cancellous bone: spongy bone; bone tissue with a lattice-like structure.
capillary: a microscopic blood vessel that connects an arteriole and a venule; the functional unit of the circulatory system.
carapace: protective covering over the dorsal part of the body of certain crustaceans and turtles.
carcinogenic: stimulating or causing the growth of a malignant tumor, or cancer.
carnivore: any animal that feeds upon another; specifically, flesh-eating mammal.
carpus: pertaining to the wrist; the eight human wrist bones.
carrying capacity: the maximum number of organisms of a species that can be maintained indefinitely in an ecosystem.
cartilage: a type of connective tissue with a solid flexible matrix.
caudal: referring to a position more toward the tail.
cell: the structural and functional unit of an organism; the smallest structure capable of performing all the functions necessary for life.
cellular respiration: the reactions of glycolysis, Krebs cycle, and electron transport system that provide cellular energy and accompanying reactions to produce ATP.
central nervous system (CNS): the brain and the spinal cord.
centriole: an organelle usually located in the centrosome, considered to be the active division center of the animal cell.
centromere: a portion of the chromosome to which a spindle fiber attaches during mitosis or meiosis.
centrosome: a dense body near the nucleus of a cell that contains a pair of centrioles.
cephalothorax: fusion of the head and thoracic regions; characteristics of certain arthropods.
cercaria: a larva of trematodes (flukes).
cerebellum: the portion of the brain concerned with the coordination of movements and equilibrium.
cerebrospinal fluid: a fluid that buoys and cushions the central nervous system.
cerebrum: the largest portion of the brain, composed of the right and left hemispheres.
cervical: pertaining to the neck or a neck-like portion of an organ.
chelipeds: pairs of pincer-like legs in most decapod crustaceans, adapted for seizing and crushing.
chitin: strong, flexible polysaccharide forming the exoskeleton of arthropods.
cholesterol: a lipid used in the synthesis of steroid hormones.
chondrocyte: a cartilage cell.
chorion: An extraembryonic membrane that participates in the formation of the placenta.
choroid: the vascular, pigmented middle layer of the wall of the eye.
chromatin: threadlike network of DNA and proteins within the nucleus.
chromosome: structure in the nucleus that contains the genes for genetic expression.
chyme: the mass of partially digested food that passes from the stomach into the duodenum of the small intestine.
cilia: microscopic, hairlike processes that move in an oar-like manner on the exposed surfaces of certain epithelial cells.
ciliary body: a portion of the choroid layer of the eye that secretes aqueous humor and contains the ciliary muscle.
ciliates: protozoans that move by means of cilia.
circadian rhythm: a daily physiological or behavioral event occurring on an approximate 24 hour cycle.
circumduction: a cone-like movement of a body part, such that the distal end moves in a circle while the proximal portion remains relatively stable.
clitoris: a small, erectile structure in the vulva of the female.
cloaca: terminal portion of the digestive tract of many animals that also may serve the excretory, reproductive, and respiratory systems.
cochlea: the spiral portion of the inner ear that contains the spiral organ (organ of Corti).
clone: asexually produced organisms having consistent genetic constitution.
cnidarian: small aquatic organisms having radial symmetry and stinging cells with nematocysts.
cocoon: protective, or resting, stage of development in certain invertebrate animals.
codon: a "triplet" of three nucleotides in mRNA that directs the placement of an amino acid into a polypeptide chain.
coelom: a fluid-filled space lined with peritoneum in visceral cavity of many bilateral animals.
collar cells: flagella-supporting cells in the inner layer of the wall of sponges.
colon: the large intestine.
common bile duct: a tube that is formed by the union of the hepatic duct and cystic duct; transports bile to the duodenum.
compact bone: tightly packed bone that is superficial to spongy bone; also called dense bone.
competition: interaction between individuals of the same or different species for a mutually necessary resource.
compound eye: arthropod eye consisting of multiple lenses.
condyle: a rounded process at the end of a long bone that forms an articulation.
conjugation: sexual union in which the nuclear material of one cell enters another cell.
connective tissue: one of the four basic tissue types within the body. It is a binding and supportive tissue with abundant matrix.
consumer: an organism that derives nutrients by feeding upon another or the remains of another.
control: a sample in an experiment that undergoes all the steps in the experiment except the one being investigated.
coral: a cnidarian that has a calcium carbonate skeleton whose remains contribute to form reefs.
cornea: the transparent, convex, anterior portion of the outer layer of the eye.
cortex: the outer layer of an organ such as the convoluted cerebrum, adrenal gland, or kidney.
costal cartilage: the cartilage that connects the ribs to the sternum.
cranial: pertaining to the cranium.
cranial nerve: one of twelve pairs of nerves that arise from the inferior surface of the brain.

cranium: the bones of the skull that enclose the brain and support the organs of sight, hearing, and balance.

crossing over: the exchange of corresponding chromatid segments of genetic material of homologous chromosomes during synapsis of meiosis I.

cyanobacteria: photosynthetic prokaryotes that have chlorophyll and release oxygen.

cytokinesis: division of the cellular cytoplasm.

cytology: the science dealing with the study of cells.

cytoplasm: the protoplasm of a cell located outside of the nucleus.

cytoskeleton: protein filaments throughout the cytoplasm of certain cells that help maintain the cell shape.

D

dendrite: a nerve cell process that transmits impulses toward a neuron cell body.

denitrifying bacteria: single-cellular organisms that convert nitrate to atmospheric nitrogen.

dentin: the principal substance of a tooth, covered by enamel over the crown and by cementum on the root.

dermis: the second, or deep, layer of skin beneath the epidermis.

descending colon: the segment of the large intestine that descends on the left side from the level of the spleen to the level of the left iliac crest.

diaphragm: a flat dome of muscle and connective tissue that separates the thoracic and abdominal cavities in mammals.

diaphysis: the shaft of a long bone.

diarthrosis: a freely movable joint.

diastole: the sequence of the cardiac cycle during which the ventricular heart chamber wall is relaxed.

diffusion: movement of molecules from an area of greater concentration to an area of lesser concentration.

dihybrid cross: a breeding experiment in which parental varieties differing in two traits are mated.

dimorphism: occurrence of two distinct forms within a species, with regard to size, color, organ structure, and so on.

diphyodont: two sets of teeth, deciduous and permanent.

diploid: a cell with two sets of chromosomes.

distal: away from the midline or origin; the opposite of proximal.

dominant: a hereditary characteristic that expresses itself even when the genotype is heterozygous.

dorsal: pertaining to the back or posterior portion of a body part; the opposite of ventral.

double helix: a double spiral used to describe the three-dimensional shape of DNA.

ductus deferens: a tube that carries spermatozoa from the epididymis to the ejaculatory duct; also called the vas deferens or seminal duct.

duodenum: the first portion of the small intestine.

dura mater: the outermost meninx (fibrous membrane) covering the central nervous system.

E

eccrine gland: a type of sweat gland that functions in evaporative cooling.

ecology: the study of the relationship of organisms with the physical environment.

ecosystem: a biological community and its associated abiotic environment.

ectoderm: the outermost of the three primary, embryonic germ layers, which gives rise to skin and nervous tissue.

edema: an excessive retention of fluid in tissues.

effector: an organ such as a gland or muscle that responds to motor stimulation.

efferent: conveying away from the center of an organ or structure.

ejaculation: the discharge of semen from the male urethra during climax.

electrocardiogram: a recording of the electrical activity that accompanies the cardiac cycle; also called ECG or EKG.

electroencephalogram: a recording of the brain wave pattern; also called EEG.

electrolyte: a solution that conducts electricity by means of charged ions.

electromyogram: a recording of the activity of a muscle during contraction: also called EMG.

electron: the unit of negative electricity.

element: a structure composed of only one type of atom (e.g., carbon, hydrogen, oxygen).

emulsification: the process of dispersing one liquid in another.

enamel: the outer, dense substance covering the crown of a tooth.

endocardium: the fibrous lining of the heart chambers and valves.

endochondral bone: bones that form as hyaline cartilage models first and then are ossified.

endocrine gland: a hormone-producing gland that secretes directly into the blood or body fluids.

endoderm: the innermost of the three primary germ layers of an embryo, which gives rise to the digestive system.

endometrium: the inner lining of the uterus.

endoskeleton: hardened, supportive internal tissue of echinoderms and vertebrates.

endothelium: the layer of epithelial tissue that forms the thin inner lining of blood vessels and heart chambers.

enzyme: a protein catalyst that activates a specific reaction.

eosinophil: a type of white blood cell that becomes stained by acidic eosin dye; constitutes about 2%–4% of the human white blood cells.

epicardium: the thin, outer layer of the heart; also called the visceral pericardium.

epidermis: the outermost layer of the skin, composed of stratified squamous epithelium.

epididymis: a coiled tube located along the posterior border of the testis; stores spermatozoa and discharges them during ejaculation.

epidural space: a space between the spinal dura mater and the bone of the vertebral canal.

epiglottis: a cartilaginous leaflike structure positioned on top of the larynx that covers the glottis during swallowing in mammals.

epinephrine: a hormone secreted from the adrenal medulla resulting in actions similar to those from sympathetic nervous system stimulation; also called adrenaline.

epiphyseal plate: a cartilaginous layer located between the epiphysis and diaphysis of a long bone. Functions in longitudinal bone growth.

epiphysis: the end segment of a long bone, distinct in early life but later becoming part of the larger bone.

epithelial tissue: one of the four basic tissue types; the type of tissue that covers or lines all exposed body surfaces.

erection: a response within an organ, such as the penis, when it becomes turgid and erect as opposed to being flaccid.

erythrocytes: red blood cells.

esophagus: a tubular organ of the GI tract that leads from the pharynx to the stomach.

estrogen: female sex hormone secreted from the ovarian (Graafian) follicle.

estuary: a zone of mixing between freshwater and seawater.

eukaryotic: possessing a nucleus and other membranous organelles characteristic of complex cells.

eustachian canal: see *auditory tube*.

evolution: organic evolution is any genetic change in organisms over time, or more precisely a change in gene frequency from one generation to another.

excretion: discharging waste material.

exocrine gland: a gland that secretes its product to an epithelial surface, directly or through ducts.

exoskeleton: an outer, hardened supporting structure secreted by ectoderm or epidermis.

expiration: the process of expelling air from the lungs through breathing out; also called exhalation.

extension: a movement that increases the angle between two bones of a joint.

external ear: the outer portion of the ear, consisting of the auricle (pinna), external auditory canal, and tympanum.

external nares: the opening into the nasal cavity; also called nostrils.

extracellular: outside a cell or cells.

extraembryonic membranes: membranes that are not a part of the embryo but are essential for the health and development of the organism.

extrinsic: pertaining to an outside or external origin.

F

facet: a small, smooth surface of a bone where articulation occurs.

facilitated transport: transfer of a particle into or out of a cell along a concentration gradient by a process requiring a carrier.

fallopian tube: see *uterine tube*.

false vocal cords: the supporting folds of tissue for the true vocal cords within the larynx.

fascia: a tough sheet of fibrous connective tissue binding the skin to underlying muscles or supporting and separating muscle.

fasciculus: a bundle of muscle or nerve fibers.

feces: waste material expelled from the GI tract during defecation, composed of food residue, bacteria, and secretions; also called stool.

fetus: the unborn offspring during the last stage of prenatal development.

filter feeder: an animal that obtains food by straining it from the water.

filtration: the passage of a liquid through a filter or a membrane.

fimbriae: fringe-like extensions from the open end of the uterine tube.

fissure: a groove or narrow cleft that separates two parts of an organ.

flagella: long slender locomotor processes characteristic of flagellate protozoans, certain bacteria, and sperm.

flexion: a movement that decreases the angle between two bones of a joint; opposite of extension.

fluke: a parasitic flatworm within the class Trematoda.

follicle: the portion of the ovary that produces the egg and the female sex hormone estrogen; the depression that supports and develops a feather or hair.

fontanel: a membranous-covered region on the skull of a fetus or baby where ossification has not yet occurred; also called a soft spot.

food web: the food links between populations in a community.

foot: the terminal portion of the lower extremity, consisting of the tarsus, metatarsus, and phalanges; a supporting structure used for locomotion.

foramen: an opening in a bone for the passage of a blood vessel or a nerve.

foramen ovale: the opening through the interatrial septum of the fetal heart; also a foramen in the floor of the sphenoid bone of the skull.

fossa: a depressed area, usually on a bone.

fossil: any preserved remains or impressions of an organism within the Earth's crust.

fourth ventricle: a cavity within the brain containing cerebrospinal fluid.

fovea centralis: a depression on the macula lutea of the eye where only cones are located, which is the area of keenest vision.

G

gallbladder: a pouchlike organ, attached to the inferior side of the liver, that stores and concentrates bile.

gamete: a haploid sex cell.

gamma globulins: protein substances that act as antibodies often found in immune serums.

ganglion: an aggregation of nerve cell bodies outside the central nervous system.

gastrointestinal tract: the tubular portion of the digestive system that includes the stomach and the small and large intestines; also called the GI tract or alimentary canal.

gene: one of the biologic units of heredity; parts of the DNA molecule located in a definite position on a certain chromosome.

gene pool: the total of all the genes of the individuals in a population.

genetic drift: evolution by chance process.

genetics: the study of heredity.

genotype: the genetic makeup of an organism.

genus: the taxonomic category above species and below family.

gill: a gas-exchange organ characteristic of fishes and other aquatic or semiaquatic animals.
gingiva: the fleshy covering over the mandible and maxilla through which the teeth protrude within the mouth; also called the gum.
gland: an organ that produces a specific substance or secretion.
glans penis: the enlarged, distal end of the penis.
glomerular capsule: the double-walled proximal portion of a renal tubule that encloses the glomerulus of a nephron; also called Bowman's capsule.
glomerulus: a coiled tuft of capillaries that is surrounded by the glomerular capsule and filters urine from the blood.
glottis: a slitlike opening into the larynx, positioned between the true vocal cords.
glycogen: the principal storage carbohydrate in animals. It is stored primarily in the liver and is made available as glucose when needed by the body cells.
goblet cell: a unicellular gland within columnar epithelia that secretes mucus.
gonad: a reproductive organ, testis or ovary, that produces gametes and sex hormones.
gray matter: the portion of the central nervous system that is composed of nonmyelinated nervous tissue.
grazer: animals that feed on low-growing vegetation, such as grasses.
greater omentum: a double-layered serous membrane that originates on the greater curvature of the stomach and extends over the abdominal viscera.
gut: pertaining to the intestine; generally a developmental term.
gyrus: a convoluted elevation or ridge.

H

habitat: the ecological abode of a plant or animal species.
hair: an epidermal structure consisting of keratinized dead cells that have been pushed up from a dividing basal layer.
hair cells: specialized receptor nerve endings for responding to sensations, such as in the spiral organ of the inner ear.
hair follicle: a tubular depression in the skin in which a hair develops.
hand: the terminal portion of the upper extremity, consisting of the carpus, metacarpus, and phalanges.
haploid: a cell with one set of chromosomes.
hard palate: the bony partition between the oral and nasal cavities, formed by the maxillae and palatine bones.
haustra: sacculations or pouches of the colon.
Haversian system: see *osteon*.
heart: a muscular, pumping organ positioned in the thoracic cavity.
hematocrit: the volume percentage of red blood cells in whole blood.
hemoglobin: the pigment of red blood cells that transports O_2 and CO_2.
hemopoiesis: production of red blood cells.
hepatic portal circulation: the return of venous blood from the digestive organs through a capillary network within the liver before draining into the heart.
herbivore: an organism that feeds exclusively on plants.
heredity: the transmission of certain characteristics, or traits, from parents to offspring, via the genes.
heterodont: having teeth differentiated into incisors, canines, premolars, and molars for specific functions.
heterotroph: an organism that utilizes preformed food.
heterozygous: having two different alleles (e.g., Bb) for a given trait.
hiatus: an opening or fissure.
hilum: a concave or depressed area where vessels or nerves enter or exit an organ.
histology: microscopic anatomy of the structure and function of tissues.
homeostasis: self-regulation of body functions to maintain an internal steady state.
homologous: similar in developmental origin and sharing a common ancestry.
hormone: a chemical substance that is produced in an endocrine gland and secreted into the bloodstream to cause an effect in a specific target organ.
host: an organism on or in which another organism lives.
hyaline cartilage: the most common kind of cartilage in the body, occurring at the articular ends of bones, in the trachea, and within the nose, which forms the precursor to most of the bones of the skeleton.
hybrid: an offspring from the crossing of genetically different strains or species.
hymen: a developmental remnant of membranous tissue that partially covers the vaginal opening.
hyperextension: extension beyond the normal anatomical position of 180°.
hypothalamus: a structure within the brain below the thalamus that functions as an autonomic center and regulates the pituitary gland.
hypothesis: a theory that is capable of explaining data and that may be used to predict the outcome of future experimentation.
hypotonic solution: a fluid environment that has a greater concentration of water and a lesser concentration of solute than the cell.

I

ileocaecal valve: a specialization of the mucosa at the junction of the small and large intestine that forms a one-way passage and prevents the backflow of food materials.
ileum: the terminal portion of the small intestine between the jejunum and caecum.
imprinting: a type of learned behavior during a limited critical period.
indigenous: organisms that are native to a particular region; not introduced.
inferior vena cava: a systemic vein that collects blood from the body regions inferior (posterior) to the level of the heart and returns it to the right atrium.
inguinal: pertaining to the groin region.
inguinal canal: the circular passage through which a testis descends into the scrotum.
insertion: the more movable attachment of a muscle, usually more distal in location.

inspiration: the act of breathing air into the alveoli of the lungs; also called inhalation.
instar: stage of insect or other arthropod development between molts.
integument: pertaining to the skin.
internal ear: the innermost portion or chamber of the ear, containing the cochlea and the vestibular organs.
internal nares: the two posterior openings from the nasal cavity into the nasopharynx; also called the choanae.
interstitial: pertaining to spaces or structures between the functioning active tissue of any organ.
intervertebral disk: a pad of fibrocartilage between the bodies of adjacent vertebrae.
intestinal gland: a simple tubular digestive gland that opens onto the surface of the intestinal mucosa and secretes digestive enzymes; also called crypt of Lieberkuhn.
intracellular: within the cell itself.
intrinsic: situated or pertaining to internal origin.
invertebrate: an animal that lacks a vertebral column.
iris: the pigmented muscular portion of the eye that surounds the pupil and regulates its diameter.
islets of Langerhans: see *pancreatic islets*.
isotope: a chemical element that has the same atomic number as another but a different atomic weight.
isthmus: a narrow neck or portion of tissue connecting two structures.

J

jejunum: the middle portion of the small intestine, located between the duodenum and the ileum.
joint capsule: a fibrous tissue cuff surrounding a movable joint.
jugular: pertaining to the veins of the neck that drain the areas supplied by the carotid arteries.

K

karyotype: the arrangement of chromosomes that is characteristic of the species or of a certain individual.
keratin: an insoluble protein present in the epidermis and in epidermal derivatives such as scales, feathers, hair, and nails.
kidney: one of the paired organs of the urinary system that contain nephrons and filter urine from the blood.

L

labia majora: a portion of the external genitalia of a female, consisting of two longitudinal folds of skin extending downward and backward from the mons pubis.
labia minora: two small folds of skin, devoid of hair and sweat glands, lying between the labia majora of the external genitalia of a female.
lacrimal gland: a tear-secreting gland, located on the superior lateral portion of the eyeball underneath the upper eyelid.
lactation: the production and secretion of milk by the mammary glands.
lacteal: a small lymphatic duct within a villus of the small intestine.
lacuna: a hollow chamber that houses an osteocyte in mature bone tissue or a chondrocyte in cartilage tissue.
lamella: a concentric ring of matrix surrounding the central canal in an osteon of mature bone tissue.
large intestine: the last major portion of the GI tract, consisting of the cecum, colon, rectum, and anal canal.
larva: an immature, developmental stage that is quite different from the adult.
larynx: the structure located between the pharynx and trachea that houses the vocal cords; commonly called the voice box.
lens: a transparent refractive structure of the eye, derived from ectoderm and positioned posterior to the pupil and iris.
leukocyte: a white blood cell; also spelled leucocyte.
ligament: a fibrous band of connective tissue that binds bone to bone to strengthen and provide support to the joint; also may support viscera.
limbic system: a portion of the brain concerned with emotions and autonomic activity.
linea alba: a fibrous band extending down the anterior medial portion of the abdominal wall.
locus: the specific location or site of a gene within the chromosome.
lumbar: pertaining to the region of the loins.
lumen: the space within a tubular structure through which a substance passes.
lung: one of the two major organs of respiration within the thoracic cavity.
lymph: a clear fluid that flows through lymphatic vessels.
lymph node: a small, oval mass located along the course of lymph vessels.
lymphocyte: a type of white blood cell characterized by a nongranular cytoplasm.

M

macula lutea: a depression in the retina that contains the fovea centralis, the area of keenest vision.
malignant: a disorder that becomes worse and eventually may cause death, as in cancer.
malnutrition: any abnormal assimilation of food; receiving insufficient nutrients.
mammary gland: the gland of the mammalian female breast responsible for lactation and nourishment of the young.
mantle: fleshy fold that envelops the viscera of a mollusk and secretes the shell.
marrow: the soft vascular tissue that occupies the inner cavity of certain bones and produces blood cells.
matrix: the intercellular substance of a tissue.
meatus: an opening or passageway into a structure.
mediastinum: the space in the center of the thorax between the two pleural cavities.
medulla: the center portion of an organ.
medulla oblongata: a portion of the brainstem between the pons and the spinal cord.
medullary cavity: the hollow center of the diaphysis of a long bone, occupied by marrow.
meiosis: nuclear reduction division by which gametes, or haploid sex cells, are formed after cytokinesis.
melanocyte: a pigment-producing cell in the deepest epidermal layer of the skin.

membranous bone: bone that forms from membranous connective tissue rather than from cartilage.
menarche: the first menstrual discharge in the human female.
meninges: a group of three fibrous membranes that covers the central nervous system.
menisci: wedge-shaped cartilage in certain movable joints.
menopause: the cessation of menstrual periods in the human female.
menses: the monthly flow of blood from the human female genital tract.
menstrual cycle: the rhythmic female reproductive cycle, characterized by changes in hormone levels and physical changes in the uterine lining.
menstruation: the discharge of blood and tissue from the uterus at the end of the human female menstrual cycle.
mesentery: a fold of peritoneal membrane that attaches an abdominal organ to the abdominal wall.
mesoderm: the middle of the three germ layers, which gives rise to bone, muscle, blood, etc.
mesothelium: a simple squamous epithelial tissue that lines body cavities and covers visceral organs; also called serosa.
metabolism: the chemical changes that occur within a cell.
metacarpus: the region of the hand between the wrist and the phalanges, including the five bones that constitute the palm of the hand.
metamorphosis: change in morphologic form, such as when an insect larva develops into the adult or a tadpole develops into an adult frog.
metastasis: the spread of a disease from one organ or body part to another.
metatarsus: the region of the foot between the ankle and the phalanges, consisting of five bones.
microbiology: the science dealing with microscopic organisms, including bacteria, fungi, viruses, and protozoa.
microvilli: microscopic, hairlike projections of cell membranes on certain epithelial cells.
midbrain: the portion of the brain between the pons and the forebrain.
middle ear: the middle of the three ear chambers, containing the three ear ossicles, in mammals.
migration: movement of organisms from one geographical site to another.
mimicry: a protective resemblance of an organism to another.
mitosis: the process of nuclear division, in which the two daughter cells are identical and contain the same number of nuclear chromosomes, typically followed by cytokinesis to form two identical daughter cells.
mitral valve: the left atrioventricular heart valve; also called the bicuspid valve.
mixed nerve: a nerve containing both motor and sensory nerve fibers.
molecule: a minute mass of matter, composed of a combination of atoms that form a given chemical substance or compound.
molting: periodic shedding of an epidermal-derived structure.
motor neuron: a nerve cell that conducts action potential away from the central nervous system and innervates effector organs (muscles and glands); also called efferent neuron.
motor unit: a single motor neuron and the muscle fibers it innervates.
mucosa: a mucous membrane that lines cavities and tracts opening to the exterior.
muscle: an organ adapted to contract; three types of muscle tissue are cardiac, smooth, and skeletal.
mutation: a variation in an inheritable characteristic, a permanent transmissible change in which the offspring differ from the parents.
mutualism: a beneficial relationship between two organisms of different species.
myelin: a lipoprotein material that forms a sheath-like covering around nerve fibers.
myocardium: the cardiac muscle layer of the heart.
myofibril: a bundle of contractile fibers within muscle cells.
myoneural junction: the site of contact between an axon of a motor neuron and a muscle fiber.
myosin: a thick filament protein that together with actin causes muscle contraction.

N

nail: a hardened, keratinized plate that develops from the epidermis and forms a protective covering on the dorsal surfaces of the digits.
nasal cavity: a mucosa-lined space above the oral cavity that is divided by a nasal septum and is the first chamber of the respiratory system.
nasal septum: a bony and cartilaginous partition that separates the nasal cavity into two portions.
natural selection: the evolutionary mechanism by which better adapted organisms are favored to reproduce and pass on their genes to the next generation.
nephron: the functional unit of the kidney, consisting of a glomerulus, glomerular capsule, convoluted tubules, and the loop of the nephron.
nerve: a bundle of nerve fibers outside the central nervous system.
neurofibril node: a gap in the myelin sheath of a nerve fiber; also called the node of Ranvier.
neuroglia: specialized supportive cells of the central nervous system.
neurolemmocyte: a specialized neuroglia cell that surrounds an axon fiber of a peripheral nerve and forms the neurilemmal sheath; also called the Schwann cell.
neuron: the structural and functional unit of the nervous system, composed of a cell body, dendrites, and an axon; also called a nerve cell.
neutron: a subatomic particle in the nucleus of an atom that has a weight of one atomic mass unit and carries no charge.
neutrophil: a type of phagocytic white blood cell.
niche: the position and functional role of an organism in its ecosystem.
nipple: a dark pigmented, rounded projection at the tip of the mammalian breast; openings of mammary glands.

nitrogen fixation: a process carried out by certain organisms, such as by soil bacteria, whereby free atmospheric nitrogen is converted into ammonia or organic N.

node of Ranvier: see *neurofibril node*.

notochord: a flexible rod of connective tissue providing skeletal support for swimming muscles in certain chordates or their embryos.

nucleic acid: an organic molecule composed of joined nucleotides, such as RNA and DNA.

nucleus: a spheroid membrane-bound body within a cell that contains the genetic factors of the cell.

nurse cells: specialized cells within the testes that supply nutrients to developing spermatozoa; also called sertoli cells.

O

olfactory: pertaining to the sense of smell.

oocyte: a developing egg cell.

oogenesis: the process of female gamete formation.

optic: pertaining to the eye and the sense of vision.

optic chiasma: an X-shaped structure on the inferior aspect of the brain where there is a partial crossing over of fibers in the optic nerves.

optic disk: a small region of the retina where the fibers of the ganglion neurons exit from the eyeball to form the optic nerve; also called the blind spot.

oral: pertaining to the mouth; also called buccal.

organ: a structure consisting of two or more tissues that performs a specific function.

organelle: a minute structure of a cell with a specific function.

organism: an individual living creature.

orifice: an opening into a body cavity.

origin: the place of muscle attachment onto the more stationary point or proximal bone; opposite of the insertion.

osmosis: the passage of a solvent, such as water, from a solution of lesser concentration to one of greater concentration through a semipermeable membrane

ossicle: one of the three bones of the middle ear of mammals.

osteocyte: a mature bone cell.

osteon: a group of osteocytes and concentric lamellae surrounding a central canal within bone tissue; also called a Haversian system.

oval window: see *vestibular window*

ovarian follicle: a developing ovum and its surrounding epithelial cells.

ovary: the female gonad in which ova and certain sexual hormones are produced.

oviduct: the tube that transports ova from the ovary to the uterus; also called the uterine tube or fallopian tube.

ovipositor: a structure at the posterior end of the abdomen in many female insects for laying eggs.

ovulation: the rupture of an ovarian follicle with the release of an ovum.

ovule: the female reproductive organ in a seed plant that contains the megasporangium where meiosis occurs and the female gametophyte is produced.

ovum: a secondary oocyte after ovulation but before fertilization.

P

palate: the roof of the mouth or oral cavity.

palmar: pertaining to the palm of the primate hand.

pancreas: organ in the abdominal cavity that secretes gastric juices into the GI tract and insulin and glucagon into the blood.

pancreatic islets: clusters of cells within the pancreas that forms the endocrine portion of the pancreas; also called islets of Langerhans.

papillae: small nipple-like projections.

paranasal sinus: a mucous-lined air chamber that communicates with the nasal cavity.

parasite: an organism that resides in or on another from which it derives sustenance but offers no benefits to the host.

parasympathetic: pertaining to the division of the autonomic nervous system concerned with activities that restore and conserve metabolic energy.

parathyroids: small endocrine glands that are embedded on the posterior surface of the thyroid glands and are concerned with calcium metabolism.

parietal: pertaining to a wall of an organ or cavity.

parotid gland: one of the paired salivary glands on the sides of the face over the masseter muscle.

parturition: the process of childbirth.

pathogen: any disease-producing organism.

pectoral girdle: the portion of the skeleton that supports the anterior appendages in a vertebrate.

pelvic: pertaining to the pelvis.

pelvic girdle: the portion of the skeleton by which the posterior appendages are supported.

penis: the external male genital organ, through which urine passes during urination and that transports semen to the female during coitus.

pericardium: a protective serous membrane that surrounds the heart.

perineum: the floor of the pelvis, which is the region between the anus and the scrotum in the male and between the anus and the vulva in the female.

periosteum: a fibrous connective tissue covering the surface of bone.

peripheral nervous system: the nerves and ganglia of the nervous system that lie outside of the brain and spinal cord.

peristalsis: rhythmic contractions of smooth muscle in the walls of various tubular organs that move the contents along.

peritoneum: the serous membrane that lines the abdominal cavity and covers the abdominal viscera; requisite lining of a true coelom.

phagocyte: any cell that engulfs other cells, including bacteria or small foreign particles.

phalanx, pl. **phalanges**: a bone of the finger or toe.

pharynx: the region of the GI tract and respiratory system located at the back of the oral and nasal cavities and extending to the larynx anteriorly and the esophagus posteriorly; also called the throat.

phenotype: the appearance of an organism caused by the genotype and environmental influences.

pheromone: a chemical secreted by one organism that influences the behavior of another.

photoperiodism: the response of an organism to periods of light and dark.

photosynthesis: the process of using the energy of the sun to make carbohydrate from carbon dioxide and water.

physiology: the science that deals with the study of body functions.

pia mater: the innermost meninx, which is in direct contact with the brain and spinal cord.

pineal gland: a small cone-shaped gland located in the roof of the third ventricle.

pituitary gland: a small, pea-shaped endocrine gland situated on the inferior surface of the brain that secretes a number of hormones; also called the hypophysis and commonly called the master gland.

placenta: the organ of metabolic exchange between the mother and the fetus.

plankton: aquatic, free-floating microscopic organisms.

plasma: the fluid, extracellular portion of circulating blood.

platelets: fragments of specific bone marrow cells that function in blood coagulation: also called thrombocytes.

pleural membranes: serous membranes that surround the lungs and line the thoracic cavity.

plexus: a network of interlaced nerves or vessels.

plica circulares: a deep fold within the wall of the small intestine that increases the absorptive surface area.

polypeptide: a molecule of many amino acids linked by peptide bonds.

pons: the portion of the brainstem just above the medulla oblongata and anterior to the cerebellum.

population: all the organisms of the same species in a particular location.

posterior (dorsal): toward the back or upper surface.

predation: the consumption of one organism by another.

pregnancy: a condition where a female has developing offspring in the uterus.

prenatal: the period of offspring development during pregnancy; before birth.

prey: organisms that are food for a predator.

producers: organisms that synthesize organic compounds from inorganic constituents within an ecosystem.

prokaryote: organism, such as a bacterium, that lacks an organized nucleus and specialized organelles.

proprioceptor: a sensory nerve ending that responds to changes in tension in a muscle or tendon.

prostate: a walnut-shaped gland surrounding the male urethra just below the urinary bladder that secretes an additive to seminal fluid during ejaculation.

protein: a macromolecule composed of one or several polypeptides.

proton: a subatomic particle of the atom nucleus that has a weight of one atomic mass unit and carries a positive charge; also a hydrogen ion.

proximal: closer to the midline of the body or origin of an appendage; opposite of distal.

puberty: the period of human development in which the reproductive organs become functional.

pulmonary: pertaining to the lungs.

pupil: the opening through the iris that permits light to enter the posterior cavity of the eyeball and be refracted by the lens.

R

receptor: a sense organ or a specialized end of a sensory neuron that receives stimuli from the environment.

rectum: the terminal portion of the GI tract, from the sigmoid colon to the anus.

reflex arc: the basic conduction pathway through the nervous system, consisting of a sensory neuron, an interneuron, and a motor neuron.

regeneration: regrowth of tissue or the formation of a complete organism from a portion.

renal: pertaining to the kidney.

renal corpuscle: the portion of the nephron consisting of the glomerulus and a glomerular capsule.

renal pelvis: the inner cavity of the kidney formed by the expanded ureter and into which the calyces open.

renewable resource: a commodity that is continually produced in the environment.

replication: the process of producing a duplicate; a copying or duplication, such as DNA replication.

respiration: the exchange of gases between the external environment and the cells of an organism.

rete testis: a network of ducts in the center of the testis.

retina: the inner layer of the eye, which contains the rods and cones.

retraction: the movement of a body part, such as the mandible, backward on a plane parallel with the ground; opposite of protraction.

rod: a photoreceptor in the retina of the eye that is specialized for colorless, dim light vision.

rotation: the movement of a bone around its own longitudinal axis.

rugae: the folds or ridges of the mucosa of an organ.

S

sagittal: a vertical plane through the body that divides it into right and left portions.

salivary gland: an accessory digestive gland that secretes saliva into the oral cavity.

sarcolemma: the cell membrane of a muscle fiber.

sarcomere: the portion of a skeletal muscle fiber between the two adjacent Z lines that is considered the functional unit of a myofibril.

Schwann cell: see *neurolemmocyte*.

scientific method: process that consists of hypothesis generation, observation, and experimentation and results in testable theories; method by which reproducible data are obtained.

sclera: the outer white layer of connective tissue that forms the protective covering of the eye.

scolex: attachment region of a tapeworm.

scrotum: a pouch of skin that contains the testes and their accessory organs.

Glossary of Terms

sebaceous gland: an exocrine gland of the skin that secretes sebum, an oily protective product.
semen: the secretion of the reproductive organs of the male, consisting of spermatozoa and additives.
semicircular canals: tubular channels within the inner ear that contain the receptors for equilibrium.
semilunar valve: crescent-shaped heart valves positioned at the entrances to the aorta and the pulmonary trunk.
seminal vesicles: a pair of accessory male reproductive organs lying posterior and inferior to the urinary bladder that secrete additives to spermatozoa into the ejaculatory ducts.
sensory neuron: a nerve cell that conducts an impulse from a receptor organ to the central nervous system; also called afferent neuron.
serous membrane: an epithelial and connective tissue membrane that lines body cavities and covers viscera; also called serosa.
sesamoid bone: a membranous bone formed in a tendon in response to joint stress.
sessile: organisms that lack locomotion and remain stationary, such as sponges and plants.
sigmoid colon: the S-shaped portion of the large intestine between the descending colon and the rectum.
sinoatrial node: a mass of cardiac tissue in the wall of the right atrium that initiates the cardiac cycle; the SA node; also called the pacemaker.
sinus: a cavity or hollow space within a body organ such as a bone.
skeletal muscle: a type of muscle tissue that is multinucleated, occurs in bundles, has crossbands of proteins, and contracts in a voluntary fashion.
small intestine: the portion of the GI tract between the stomach and the caecum; functions in absorption of food nutrients.
smooth muscle: a type of muscle tissue that is nonstriated, composed of fusiform and single-nucleated fibers, and contracts in an involuntary, rhythmic fashion within the walls of visceral organs.
solute: a substance dissolved in a solvent to form a solution.
solvent: a fluid such as water that dissolves solutes.
somatic: pertaining to the nonreproductive (nonvisceral) parts of the body.
species: a group of morphologically similar (common gene pool) organisms that are capable of interbreeding and producing fertile offspring.
spermatic cord: the structure of the male reproductive system composed of the ductus deferens, spermatic vessels, nerve, cremasteric muscle, and connective tissue.
spermatogenesis: the production of male sex gametes, or spermatozoa.
spermatozoan: a sperm cell, or gamete.
sphincter: a circular muscle that constricts a body opening or the lumen of a tubular structure.
spinal cord: the portion of the central nervous system that extends from the brainstem through the vertebral canal, also called "dorsal nerve cord."
spinal nerve: one of the 31 pairs of nerves that arise from the spinal cord.
spiracle: a respiratory opening in certain animals such as arthropods and sharks.
spleen: a large, blood-filled organ located in the upper left of the abdomen and attached by the mesenteries to the stomach.
spongy bone: a type of bone that contains many porous spaces; also called cancellous bone.
stomach: a pouch-like digestive organ between the esophagus and the duodenum.
submucosa: a layer of supportive connective tissue that underlies a mucous membrane.
succession: ecological stages by which a particular biotic community gradually changes until there is a community of climax vegetation.
superior vena cava: a large systemic vein that collects blood from regions of the body superior to the heart and returns it to the right atrium.
surfactant: a substance produced by the lungs that decreases the surface tension within the pulmonary alveoli.
suture: a type of immovable joint articulating between bones of the skull.
sympathetic: pertaining to that part of the autonomic nervous system concerned with processes involving the utilization of energy.
synapse: a minute space between the axon terminal of a presynaptic neuron and a dendrite of a postsynaptic neuron or an effector cell.
synovial cavity: a space between the two bones of a diarthrotic joint, filled with synovial fluid.
system: a group of body organs that function together.
systole: the muscular contraction of the ventricles of the heart during the cardiac cycle.
systolic pressure: arterial blood pressure during the ventricular systolic phase of the cardiac cycle.

T

target organ: the specific body organ that a particular hormone affects.
tarsus: the seven bones that form the ankle.
taxonomy: the science of describing, classifying, and naming organisms.
tendo calcaneus: the tendon that attaches the calf muscles to the calcaneus bone.
tendon: a band of dense regular connective tissue that attaches muscle to bone.
testis: the primary reproductive organ of a male, which produces spermatozoa and male sex hormones.
tetrapod: a four-appendaged vertebrate, such as amphibian, reptile, bird, or mammal.
thoracic: pertaining to the chest region.
thoracic duct: the major lymphatic vessel of the body, which drains lymph from the entire body except the upper right quadrant and returns it to the left subclavian vein.
thorax: the chest.
thymus gland: a bi-lobed lymphoid organ positioned in the upper mediastinum, posterior to the sternum and between the lungs.

tissue: an aggregation of two or more types of cells and their binding intercellular substance, joined to perform a specific function.

tongue: a protrusible muscular organ on the floor of the oral cavity.

trachea: the airway leading from the larynx to the bronchi; also called the windpipe.

tract: a bundle of nerve fibers within the central nervous system.

trait: a distinguishing feature studied in heredity.

transverse colon: a portion of the large intestine that extends from right to left across the abdomen between the hepatic and splenic flexures.

tricuspid valve: the heart valve between the right atrium and the right ventricle.

true vocal cords: folds of the mucous membrane in the larynx that produce sound as they are pulled taut and vibrated.

turgor pressure: osmotic pressure that provides rigidity to a cell.

tympanic membrane: the membranous eardrum, positioned between the outer and middle ear; also called the tympanum.

U

umbilical cord: a cordlike structure containing the umbilical arteries and vein that connects the fetus and the placenta.

umbilicus: the site where the umbilical cord was attached to the fetus: also called the navel.

ureter: a tube that transports urine from the kidney to the urinary bladder.

urethra: a tube that transports urine from the urinary bladder to the outside of the body.

urinary bladder: a distensible sac in the pelvic cavity that stores urine.

uterine tube: the tube through which the ovum is transported to the uterus and where fertilization takes place: also called the oviduct or fallopian tube.

uterus: a hollow, muscular organ in which a fetus develops. It is located within the female pelvis between the urinary bladder and the rectum.

uvula: a fleshy, pendulous portion of the soft palate that blocks the nasopharynx during swallowing.

V

vagina: a tubular organ that leads from the uterus to the vestibule of the female reproductive tract and receives the male penis during coitus.

vein: a blood vessel that conveys blood toward the heart.

ventral: toward the lower surface of the body.

vertebrate: an animal that possesses a vertebral column.

vestibular window: a membrane-covered opening in the bony wall between the middle and inner ear, into which the footplate of the stapes fits.

viscera: the organs within the abdominal or thoracic cavities.

vitreous humor: the transparent gel that occupies the space between the lens and retina of the eye.

vulva: the external genitalia of the female that surround the opening of the vagina; also called the pudendum.

Z

zygote: a fertilized egg cell formed by the union of a sperm and an ovum.

Index

A

A band 7
Abdomen 75, 77, 79, 80, 81, 82, 83, 88, 89, 150, 160, 165, 172, 174, 175, 177, 198
Abdominal
 aorta 162, 164, 172, 189, 190, 215, 221
 artery 81
 cavity 155, 177, 191, 198
 claw 80
 muscle 124, 126, 152, 161, 166, 167, 168, 174, 175, 176, 179, 181, 183, 184, 206, 207
 organs 171
 scales 135
 setae 80
 spine 77
 vein 113, 127, 128, 129, 130, 144
 viscera 164, 177
Abdominopelvic cavity 198
Abducens nerve 114, 194, 196
Abductor 124, 209
Acanthocephala 69
Acanthopleura granulata. See Chiton
Accessory nerve 194, 196
Acetabulum 123, 205
Acoustic meatus 173, 179, 201, 202, 203
Acromiodeltoid 161, 179, 182
Acromion 136, 204
Acromiotrapezius 175, 176, 179, 180, 182, 183
Acropora. See Coral, elkshorn
Actinodiscus. See Anemone, disk
Actinopterygii 99, 100, 107, 115
Adductor 55, 56, 110, 124, 125, 126, 167, 168, 175, 184, 185, 206, 207, 208
Adipocyte 10
Adipose tissue 199, 224
Adrenal gland 165, 173, 212, 213, 221
Agalychnis callidryas. See Frog, red-eyed tree
Agnatha 99, 107

Aix sponsa. See Duck, wood
Algae 1, 13, 40, 69
 brown vii, 21
 golden vii, 21, 22
 green vi, vii
 red vi, vii, 21
Alligator 131, 145
 American 133, 134
Alligator mississippiensis. See Alligator, American
Alveolar duct 217, 224
Amblyrhynchus cristatus. See Iguana, Galapagos marine
Ambystoma mexicanum. See Axolotl
Ambystoma tigrinum. See Salamander, tiger
Ammonite 57
Amnion 119, 131, 132, 155, 197
Amniotes 100, 131
Amniotic egg 99, 100, 131
Amoeba vi, vii, 24
Amoeba proteus 24
Amoebocytes 29
Amphibia 99, 100, 119, 120
Amphioxus 99
Amphiprion melanopus. See Clownfish, tomato
Amphiuma 120, 121
Amphiuma means. See Amphiuma
Ampulla 94, 95, 97, 224
Anal 109, 116, 117, 118, 135, 165, 191, 219
Anaphase 13, 14, 15
Anas cyanoptera. See Duck, cinnamon teal
Anconeus 124, 209
Anemone 33, 34, 40, 41
 disk 41
 sunburst 40, 41
 tube 40
Anhinga, American 148
Anhinga anhinga. See Anhinga, American
Animal cell cycle 14
Ankle 121, 160, 166, 174, 177
Annelida vii, viii, ix, 52, 61, 62, 69
Anopla 61, 68
Antebrachium 121, 159, 182, 183, 198
Antelope (pronghorn) 158

Anterior chamber 213
Anthopleura sola. See Anemone, sunburst
Anthozoa 33, 34, 40
Antilocapra americana. See Antelope (pronghorn)
Antrum 223
Anus 17, 19, 53, 54, 56, 58, 61, 65, 66, 67, 69, 71, 74, 77, 79, 80, 81, 89, 94, 95, 96, 97, 98, 102, 103, 118, 137, 138, 144, 164, 166, 191, 208, 219, 224
Aorta 56, 58, 83, 104, 106, 107, 111, 112, 113, 127, 129, 137, 144, 161, 162, 163, 164, 165, 169, 170, 172, 173, 186, 187, 189, 190, 191, 192, 215, 216, 221
Aortic arch 99, 107, 108, 138, 144, 145, 153, 154, 155, 161, 163, 177, 186, 187, 188, 189, 192, 215, 216
Apalone spinifera. See Turtle, spiny soft-shell
Apex of heart 154, 187, 192, 215, 216
Apicomplexa vii, 21, 22, 24
Aplysia californica. See Sea hare
Apocrine sweat gland 199
Appendix 219
Aqueduct 195, 210
Ara ararauna. See Macaw, blue and gold
Ara macao. See Macaw, scarlet
Arachnid 78
Arachnida 75, 76, 78
Arbacia. See Sea urchin
Arbor vitae of cerebellum 210
Archaea vi, vii, 21
Archaeopteryx 145
Archaeornithes 145
Archispirostreptus gigas. See Millipede, African giant
Ardea herodias. See Heron, great blue
Arenicola marina. See Lugworm
Areola 224
Argiope trifasciata. See Spider, orb weaver

Ariolimax californicus. See Slug, banana
Armadillidium See Pill bug
Arteriole 197, 217, 218
Artery 20, 58, 81, 89, 111, 112, 113, 127, 128, 130, 137, 138, 142, 144, 153, 154, 161, 162, 163, 164, 167, 170, 171, 172, 173, 185, 186, 187, 188, 189, 190, 191, 192, 208, 213, 215, 216, 221, 222, 223
Arthropoda vii, viii, ix, 69, 75, 76
Articulata. See Crinoid, red stalked
Ascaris 69, 70, 71, 72, 73
Ascending colon 219
Ascon 30
Asexual reproduction 13, 14, 33, 69
Aspergillus vi
Asterias. See Sea star
Asteroidea 92, 94
Atrioventricular valve 216
Atrium 56, 102, 103, 104, 106, 111, 112, 113, 122, 127, 130, 137, 142, 154, 177, 187, 188, 192, 216
Auditory
 canal 166, 214
 cerebral cortex 210
 nerve 114
 ossicle 99, 155, 214
 tube 155, 214
Aurelia 38, 39
Auricle 44, 55, 114, 130, 138, 144, 159, 163, 166, 171, 172, 192, 214, 216
Aves 99, 131, 145, 146, 147
Avocet, American 148
Axilla 198, 224
Axillary 204
 artery 154, 161, 163, 167, 186, 215
 vein 161, 186, 187, 215
Axis 160, 178, 203
Axolotl 120, 121
Axon 8, 12, 66, 211, 212
Aythya americana. See Duck, redhead
Azygos vein 163, 187

B

Bacteria vi, vii, 1, 2, 14, 21, 69, 75

Banana vi
Basale 199
Baseodiscus mexicanus. See Worm, zebra ribbon
Basilar membrane 214
Basiliscus plumifrons. See Basilisk, green or plumed
Basilisk, green or plumed 133
Basophil 6
Bat 156
 lesser long-nosed 158
 Malaysian fruit 157
Bathocyroe fosteri. See Comb jelly
Beak 58, 136, 145, 148
Bear 156
 grizzly 157
Beetle 75
 click 87
 milkweed 84
 tiger 76
 scarab 87
Beroe cucumis 34
Biceps
 brachii 152, 161, 167, 168, 206, 207
 femoris 124, 126, 166, 167, 168, 175, 176, 179, 184, 185, 206, 207, 208
Bicuspid valve 187, 192, 216
Binary fission 13, 14
Bipalium 44
Bipedal vertebrate 198
Bipolar neuron 213
Bison, American 158
Bison bison. See Bison, American
Bivalvia 51, 55
Blaberus giganteus. See Cockroach, giant
Blackbird, Brewer's 147
Blastocoel 17, 18, 19
Blastula 17, 18, 37
Blood vessel 12, 20, 63, 65, 66, 199, 213
Boa 141
Bone 7, 11, 116, 117, 122, 123, 132, 136, 143, 145, 152, 155, 160, 177, 178, 179, 197, 199, 200, 201, 202, 203, 205, 219
Bony fishes 99, 107, 108, 117
Brachial 58
 artery 139, 161, 163, 186, 215

nerve 114
plexus 167, 171, 172
vein 113, 130, 186
Brachialis 166, 168, 206, 207
Brachiocephalic
artery 163, 192, 215
muscle 167, 168
trunk 138, 163, 185, 186, 187, 188, 215, 216
vein 185, 186, 187, 188, 189, 215
Brachiopoda vii, viii, 51, 59, 61, 69, 76
Brachiopods 51, 52
Brachioradialis 182, 183, 206, 209
Brachium 79, 121, 159, 182, 183, 198
Brain 20, 43, 50, 58, 65, 66, 74, 79, 81, 83, 89, 99, 106, 107, 108, 114, 118, 119, 131, 145, 155, 173, 192, 193, 194, 195, 196, 197, 210
Branchiostoma. See Lancelet
Breast 150, 218, 222, 224
Brittle star 92
green 93
Bronchiole 217, 218
Bronchus 217, 218
Brown recluse. See Spider, brown recluse
Bryozoa vii, viii, 51, 60, 69
Bryozoan 51, 52, 60
flute 60
freshwater 60
lacy 60
marine 60
Bubo virginianus. See Owl, great horned
Budding 13, 33, 35, 48
Bufo alvarius. See Toad, Colorado River
Bufo marinus. See Toad, cane or marine
Bufo woodhousii. See Toad, Woodhouse's
Bulb 199
aquapharyngeal 97
buccal 58
flame 74
olfactory 114, 194, 195
Bulbourethral gland 164, 165, 191, 223
Bunting, lazuli 101, 148
Buteo jamaicensis. See Hawk, red-tailed
Butorides sundevalli. See Heron, lava
Butterfly 85, 88
Buttock 198

C

Caecilian 119
Cameroon 120, 121
Caecum 203
Caiman 131
Cuvier's dwarf 134
Calcaneus 124, 125, 184, 185, 206
Calcarea viii, 29, 30
Calf 198
Callipepla californica. See Quail, California
Calyx 92, 222
Cambarus. See Crayfish
Canaliculi 7
Capillary 6, 217, 218
Capitulum 204
Capsule 2, 222
Capybara 158
Cardiac muscle 12
Carotid artery 112, 113, 127, 128, 130, 137, 138, 144, 153, 163, 171, 172, 185, 187, 188, 189, 216
Carpal bone 122, 123, 136, 160, 178, 200
Cartilage 11, 58, 105, 106, 107, 109, 110, 200, 203, 217
Cartilaginous fishes 99, 107
Carybdea sivickisi. See Jellyfish, box
Cat 159, 177–191
Caudal 159, 165, 166
artery 161, 162, 186
fin 102–105, 109, 110, 116–118
muscle 175, 176, 180, 184
vein 113, 186
vena cava 128, 129, 130, 161, 163, 164, 172, 173, 189, 190
vertebra 107, 110, 122, 136, 143, 160, 178
Caudofemoralis 152, 175, 176, 179, 180, 184, 185
Caulastrea furcata. See Coral, candy cane
Celiac
artery 112, 113
trunk 161, 162, 186, 189, 215
Cell vii, ix, 1–3, 5–18, 21–26, 28–29, 31, 33–35, 45, 65, 73, 75, 108, 119, 155, 159, 197, 211–212, 214, 218, 221, 223–224
Centipede 75, 90
Florida blue 90

giant Sonoran 90
Vietnamese 90
Centriole 2, 3, 4, 13, 15
Centromere 14, 15
Centrosome 2, 3
Centruroides hentzi. See Scorpion, bark
Cephalochordata viii, 99, 100, 102
Cephalopod 56
fossilized 57
Cephalopoda 51, 56
Ceratium 23
Cerebellum 169, 173, 193, 195, 196, 210, 211
Cerebral 168, 169, 170, 185
aqueduct 194, 195, 210
arterial circle 194
cortex 185, 210
fissure 193
ganglion 47, 54, 61, 89
hemisphere 193
peduncle 194, 195, 210
Cerebratulus lacteus. See Worm, milky ribbon
Cerebrum 145, 155, 169, 173, 193, 195, 210
Cervical
muscle 163
region 198
trunk 163
vertebra 99, 136, 152, 155, 160, 178, 200, 203
Cervix 26, 165, 222, 224
Cestoda 43, 48
Chamaeleo calyptratus. See Chameleon, veiled
Chameleon
panther 133
veiled vi
Chelonia mydas. See Turtle, Hawaiian green
Chelonia mydas agassisi. See Turtle, Galapagos green sea
Chelonoidis nigra. See Tortoise, Galapagos
Chick development 20
Chilopoda 75, 76, 90
Chimaera 100, 107, 108
Chimpanzee 101, 156
Chiton 51, 52, 53
Choanocytes 29, 31
Chondrichthyes 99, 100, 107, 108
Chondrocyte 11
Chordae tendineae 187, 192

Chordata vii, viii, ix, 91, 99, 101
Chordeiles minor. See Nighthawk
Choroid 210, 213, 214
Chromatid 13, 14, 15
Chromatin strand 14
Chromosome vii, 1, 13, 14, 15
Chrysaora colorata. See Jelly, purple-striped
Chrysaora fuscescens. See Sea nettle
Chrysaora melanaster. See Jellyfish, red-striped
Chrysophyte 23
Cicada 84
Cicindela fulgida. See Beetle, tiger
Ciliary body 213, 214
Ciliates vi, vii, 21, 22, 27
Ciona intestinalis. See Tunicate
Clam vii, ix, 23, 51, 55, 56, 91
giant 23, 52, 55
Clavicle 123, 160, 200, 218
Clitellata 61, 64, 67
Clitoris 164, 190, 224
Clonorchis. See Fluke, liver
Clonorchis sinesis. See Fluke, human liver
Clownfish, tomato 116
Cnidaria vii, viii, ix, 33, 34, 35
Cnidocytes ix, 33, 35
Cobra, spitting 143
Cochlea 214
Cochlear duct 214
Cockroach, giant 84
Colaptes auratus cafer. See Flicker, red-shafted or northern
Colibri coruscans. See Hummingbird, sparkling violetear
Collagenous fiber 8, 10
Collar cells 29, 31
Colliculus 195, 196, 210
Colobocentrotus atratus. See Sea urchin, helmet
Colon 122, 138, 142, 144, 164, 170, 172, 176, 177, 189, 190, 219, 220
Colony 22, 36, 37, 38, 40
Columba. See Pigeon, rock dove

Comb jelly 34
Arctic 34
warty 34
Common ancestor 100
Concha 201, 202
Cone 213, 214
Connective tissue 1, 8, 10, 12, 160, 165, 174, 177, 224
Conolophus subcristatus. See Iguana, Galapagos land
Coral vii, ix, 33, 34, 40, 42, 140
brain 42
candy cane 42
elkshorn 42
firecracker 40
mushroom 42
Cornea 213, 214
Corona 69, 74, 223
Corpus 155, 195, 196, 210, 223
Costal 136
cartilage 200, 203
scales 135
Crab 68, 75
hermit 80
horseshoe 75, 77
Sally Lightfoot 80
Cranial 106, 108, 114, 119, 130, 131, 159, 161, 162, 163, 164, 166, 172, 175, 176, 185, 187, 188, 189, 192, 197, 203
Crayfish 81, 82, 83
Cricket, house 85
Crinoid 91, 92
red stalked 93
Crinoidea 92
Crocodile 100, 131, 143, 145
Johnston's freshwater 134
Crocodylus johnstoni. See Crocodile, Johnston's freshwater
Crotaphatrema bornmuelleri. See Caecilian, Cameroon
Crus 164, 191, 210
Ctenocephalide. See Flea
Ctenophora vii, viii, 33, 34
Cubital 186, 198, 215
Cubozoa 33, 34, 40
Cucumaria. See Sea cucumber
Curvature 220
Cutaneous 220
Cuttlefish 52, 56
Cynomys parvidens. See Prairie dog, Utah
Cytokinesis 13, 14, 15

Index

Cytoplasm 1, 2, 3, 6, 8, 13, 21, 23, 24, 211
Cytoskeleton vii, 2, 3, 22

D

Daphnia. See Water flea
Daughter cells 13, 15
Deer, mule 157
Deltoid 124, 125, 166, 168, 198, 204, 206, 207
Demospongiae 29, 32
Dendrite 7, 8, 211
Dendroaspis jamesoni. See Snake, Jameson's mamba
Denticle 54
Dentin 107, 219
Dermanyssus gallinae. See Mite, red
Diadophis punctatus. See Snake, ring-neck
Diaphragm 99, 155, 162, 163, 164, 169, 170, 171, 172, 177, 185, 186, 198, 215, 217, 218, 220, 221
Diatoms vi, vii, 21, 22
Diceroprocta apache. See Cicada
Diceros bicornis. See Rhinoceros, black or hook-lipped
Digastric 167, 168, 181
Digestive system 89, 91, 107, 197, 219
Dinoflagellates vi, vii, 21; 22, 23
Diploid 16, 28
Diplopoda 75, 76, 90
Dipsochelys dussumieri. See Tortoise, Aldabra giant
Dirofilaria immitis. See Heartworm
Disk 8, 41, 75, 83, 92, 121, 205
Dissection iii, 160, 165, 174, 177, 192
DNA vi, 1, 2, 13, 14, 28, 69
Dolphin, bottlenose 155–157
Dove 146
  mourning 146–147
  rock 151
Dragonfly, flame skimmer 84
Dromaius novaehollandiae. See Emu
Duck 146
  cinnamon teal 13
  redhead 146–147
  wood 149, 151
Ductus (Vas) deferens 45, 47, 48, 49, 65, 71, 72, 73, 83, 142, 144, 164, 165, 173
Dugesia 44, 45
Duodenum 122, 128, 144, 177, 219, 220
Dura mater 193, 194, 210

E

Eagle, American bald 148
Ear 20, 140, 145, 155, 165, 197, 214
Earthworm 61, 62, 64, 65, 66, 121
Ecdysis 85, 140
Ecdysozoa 69
Echinarachnius parma. See Sand dollar, common
Echinoderm 13, 14, 91, 93, 99
Echinodermata vii, viii, ix, 91, 92, 100
Echinoidea 92, 96
Ectoderm 17, 19, 36, 43
Eel, vinegar 70
Egg 13, 16, 17, 18, 29, 33, 35, 36, 37, 38, 39, 43, 45, 46, 48, 50, 60, 64, 65, 73, 75, 79, 80, 84, 85, 91, 99, 100, 119, 131, 141, 142, 144, 145, 151, 155
Emu 146–147
Enamel 107, 219
Endocrine 197, 212
Endoderm 17, 19, 36, 43, 45
Enterobius vermicularis. See Pinworm
Enteroctopus. See Octopus, giant
Enteropneusta 92, 98
Entoprocta vii, viii, 69
Epididymis 164, 165, 173, 223
Epiglottis 188, 217
Epinephelus lanceolatus. See Grouper, giant
Epithelial tissue 1, 8
Epithelium 9, 10, 61, 65, 66, 217, 218, 221, 223
Epitrochlearis 175, 182, 183
Eremobates pallipes. See Solpugid
Erythrocyte 6, 25, 155
Esophagus 45, 47, 56, 58, 65, 66, 83, 89, 97, 102, 103, 106, 111, 118, 122, 137, 138, 144, 152, 153, 154, 162, 163, 188, 217, 219, 220
Euglena vi, 26
Eukarya vi, vii
Eukaryotic cells 1, 21
Euphagus cyanocephalus. See Blackbird, Brewer's
Euspongia 29
Eutardigrade 90
Extension 2, 24, 211
Extensor carpi radialis 166, 167, 168, 182, 183, 209
Extensor carpi ulnaris 124, 168, 183, 206, 209
Extensor digitorum 124, 166, 167, 168, 183, 206, 209
External abdominal oblique 124, 125, 126, 152, 161, 166, 167, 168, 174, 175, 176, 179, 183, 184, 206, 207
External intercostal 168, 207
Eye 19, 20, 38, 50, 51, 54, 58, 59, 77, 79, 81, 87, 105, 106, 109, 114, 116, 118, 121, 122, 132, 135, 143, 145, 155, 173, 197, 210, 213, 214
  compound 75, 77, 80, 81, 82, 83, 86, 88, 89
Eye worm 70
Eyelid 140, 155, 159, 166
Eyespot 26, 44, 74, 94

F

Facial nerve 110, 114, 163, 194
Falciform ligament 220
Falco peregrinus. See Falcon, peregrine
Falco sparverius. See Falcon, American kestrel
Falcon 146
  American kestrel 149
  peregrine 147
Fascia 166, 167, 168, 174, 175, 176, 179, 180, 184, 199, 206, 208, 223
Fasciola hepatica. See Fluke, sheep liver
Feather 92, 99, 145, 150, 151
Felis rufus. See Bobcat
Femoral
  artery 113, 130, 173, 186, 208, 215
  pores 142
  region 198, 207, 208
  scales 135
  vein 113, 130, 186, 208, 215
Femur 79, 86, 88, 122, 123, 136, 152, 160, 178, 200, 205
Fertilization 13, 16, 17, 23, 25, 28, 29, 107, 108, 119, 131, 145, 155
Fetal pig 165–173
Fibroblast 8, 10
Fibrocartilage 11
Fibula 122, 136, 160, 178, 200, 205
Fimbria 191, 224
Fission 13, 14, 27, 28
Flagellated protozoans 26
Flame cells ix, 43
Flamingo 146
  Chilean 146–148
Flatworm vii, ix, 13, 14, 43
Flea 87
  water 80
Flexor carpi
  radialis 167, 206, 209
  ulnaris 152, 167, 182, 206, 209
Flicker
  red-shafted or northern 147, 148
Fluke 43, 47
  cow liver 45
  human liver 47, 48
  sheep liver 45, 46
Follicle 16, 159, 199, 212, 223
Follicular
  cells 223
  fluid 223
Foramen 195, 201, 202, 203, 205, 210
Fornix 195, 196, 210
Fossa 123, 198, 202, 203, 204, 205, 206
Fovea centralis 213
Fragmentation vi, 13, 14
Frog 16, 18, 19, 60, 99, 100, 119, 123–130
  blue-webbed flying treefrog 120
  canyon tree 120
  leopard 121
  red-eyed tree 120
  white-lipped tree 121
Frontal
  bone 122, 136, 179, 200–203
  cerebrum 173, 210
  lobe 193
  plane 159, 197
  sinus 202
Furcifer pardalis. See Chameleon, panther

G

Gallbladder 122, 128, 129, 137, 138, 142, 170, 171, 185, 219, 220, 221
Gametes 13, 16, 25, 33, 91, 131, 197
Ganglia ix, 43, 54, 61
Ganglion 47, 58, 81, 89, 114, 213
Gastric 39, 74, 83, 89, 113, 127, 130, 161, 162, 164, 186, 221
Gastrocnemius 124–126, 152, 166–168, 175–176, 184–185, 206, 208
Gastropoda 51, 54
Gastrotricha vii, viii, 69
Gastrula 17, 18, 19
Gavia pacifica. See Loon, pacific
Gavialis gangeticus. See Gharial
Genitalia 160, 165, 174, 177, 223, 224
Gharial 134
Gila monster 139
Gluteus 124, 126, 166–168, 175–176, 179–180, 184–185, 206–208
Glycera americana. See Bloodworm
Glycogen granules 3
Gnathostomata 99
Goblet cell 8, 9, 218
Golden algae. See Algae, golden
Golgi complex 2, 3
Goniastrea. See Brain coral
Gopherus agassizii. See Tortoise, gopher
Gorilla, western lowland 158
Gorilla gorilla gorilla. See Gorilla, western lowland
Gracilis 124–126, 168, 174–175, 184, 196, 206–207
Grantia 29, 31
Granulosum 199
Grapsus grapsus. See Crab, Sally Lightfoot
Grasshopper 85, 88, 89
  Eastern lubber 84
Grebe 146
  eared 146–147
Grouper 116

giant 101
Guitarfish 108
Gull 146
 Franklin's 147
 ring-billed 148
Gut 19, 25, 46
 hindgut 19, 80, 89
 midgut 80, 89, 103

H

Hadogenes troglodytes. See Scorpion, flat rock
Hadrurus arizonensis. See Scorpion, Arizona hairy
Hair 86, 87, 99, 155, 159, 199, 214
Haliaeetus leucocephalus. See Eagle, American bald
Haploid 13, 16, 28
Haplopelma lividum. See Tarantula, cobalt blue
Hard palate 169, 188, 217
Haversian
 canal 11, 199
 system 199
Hawk
 nighthawk 151
 red-tailed 149
Head ix, 3, 16, 20, 43, 44, 51, 53, 54, 57, 58, 61, 72, 75, 86, 88, 89, 99, 100, 104, 105, 106, 119, 135, 140, 155, 163, 198
Heart ix, 19, 20, 54, 56, 58, 61, 65, 66, 73, 75, 79–81, 89, 99, 107, 108, 111, 112, 118, 119, 122, 127, 128, 137, 138, 142, 144, 145, 153–155, 161–163, 169, 170, 172, 176, 177, 185–189, 192, 197, 198, 215, 216, 218
Heartworm 73
Heloderma suspectum. See Gila monster
Hemichordata vii, viii, 91, 92, 98
Hemipenes 131, 132, 142–144
Hemiscolopendra marginata. See Centipede, Florida blue
Hepatic
 artery 113, 162, 186
 caecum 104
 system 164
 vein 113, 130, 144, 164, 186, 215

Heron 146
 great blue 158
 lava 147
Heterocentrotus. See Sea urchin, pencil
Heterotardigrade 90
Heterotrophic vii, 1
Hexactinellida 29
Hippodiplosia insculpta. See Bryozoan, flute
Hirudinea 61, 67
Holothuroidea 92, 97
Honeybee 85, 86
Hooded merganser 145
Hookworm 69, 70
Horn of uterus 137, 165, 190, 191
Housefly 86
Humerus 122, 123, 136, 152, 160, 178, 200, 204
Hummingbird 146, 151
 sparkling violetear 147
 broad-tailed 148, 151
Hyaline cartilage 11, 217
Hydra vii, ix, 13, 33, 34, 35, 36
Hydrochoerus hydrochaeris. See Capybara
Hydrolagus colliei. See Chimaera
Hydrozoa 33, 34, 35
Hyla arenicolor. See Frog, canyon tree
Hypodermis 199
Hypoglossal
 canal 202
 nerve 196
Hypophysis 194, 196
Hypothalamus 196, 210, 212
Hyrax 156, 157

I

Iguana
 Galapagos land ii, 139
 Galapagos marine 133
Ileum 111, 153, 162, 176
Iliac
 artery 113, 127, 129, 130, 162, 173, 186, 208, 215
 bone 205
 crest 207
 vein 113, 186, 215
Iliacus 124, 126, 167, 168
Iliolumbar

artery 161, 162, 173, 186
 muscle 124
 vein 161, 186
Ilium 123, 152, 160, 178, 198, 200, 205
Inferior vena cava 172, 173, 186, 187, 189, 190, 191, 215, 216, 221
Infraspinatus 175, 176, 206, 207
Insect vii, ix, 75, 84, 86, 87, 119, 148
Insecta 75, 76, 84
Interlobular 217, 224
Internal abdominal oblique 124, 168, 184
Interosseous margin 204
Interparietal bone 179
Interphase 13, 14
Interventricular septum 187, 192, 216
Intervertebral disk 205
Intestine 44, 45, 47, 50, 54, 56, 58, 63, 65–67, 71–74, 79, 81, 83, 96, 97, 102, 103, 106, 111, 118, 144
 large 127–129, 169, 219
 small 50, 127, 128, 137, 138, 142, 144, 154, 169, 170–172, 176, 185, 189, 190, 219, 220, 221
Iris 213, 214
Ischium 123, 136, 152, 160, 178, 200, 205
Iterus bullockii. See Oriole, Bullock's

J

Jack mackerel 115
Jaw 99, 100, 107–110, 112, 155, 165
Jejunum 162, 221
Jelly, purple-striped 39
Jellyfish vii, ix, 33, 34, 39
 box 34, 40
 red-eye 38
 red-striped 39
Joint ix, 75, 155, 197, 203, 205
Jugular 106, 113, 130, 144, 161, 163, 171, 172, 185–188, 203, 215

K

Kangaroo 156
 eastern gray 156–157

Katydid, greater arid-land 84
Kelp, giant 21, 60
Kidney 54–56, 58, 106, 107, 111, 118, 119, 127, 129, 144, 153, 164, 165, 170–173, 177, 189–191, 197, 213, 221, 222
Kingena. See Brachiopod
Kingfisher 146
 belted 147
Kinorhyncha vii, viii, 69
Komodo dragon 133

L

Labium 86, 89, 224
Lacuna 7, 11, 199
Lamella 11, 87, 199
Lamina 8, 218, 221
Lamp shell vii, 51, 59
Lamprey 99, 104, 105, 106, 107
Lampropeltis getulus. See Snake, California king
Lampropeltis pyromelana. See Snake, mountain kingsnake
Lampropeltis triangulum elapsoides. See Snake, scarlet kingsnake
Lancelet vii, ix, 99–104
Larus delawarensis. See Gull, ring-billed
Latissimus dorsi 124, 152, 161, 166, 168, 176, 179, 180–184, 206, 207
Latrodectus hesperus. See Spider, black widow
Leech 61, 62, 67
Lemur
 flying 156
 ring-tailed 158
Lemur catta. See Lemur, ring-tailed
Leptonycteris yerbabuenae. See Bat, lesser long-nosed
Leucon 32
Leuconoid sponge 32
Leucophaeus pipixcan. See Gull, Franklin's
Leucosolenia 29, 30
Leukocytes 6
Levator scapulae 176, 182
Libellula saturata. See Dragonfly, flame skimmer
Ligia italica. See Sea slater

Limb bud 19
Linea alba 110, 125, 161, 175
Lingula. See Lamp shell
Lingulata 51
Lionfish 116
Litoria infrafrenata. See Frog, white-lipped tree
Liver 19, 25, 45–48, 54, 104, 106, 111, 118, 122, 127, 128, 137, 138, 142, 144, 153, 154, 162, 169–172, 176, 177, 185, 189, 197, 215, 218–221
Lizard 78, 131, 132, 140, 143
 argus monitor 141
 desert spiny 133
 eastern glass 139
 fence 142
 horned 139
 savannah monitor 80, 140, 141
Loa loa. See Eye worm
Loligo. See Squid
Longitudinal 160, 174, 177
 cerebral fissure 193
 fissure 173
 muscle ix, 44, 61, 63, 65, 66, 69, 73, 221
 nerve 48
Loon 146
 pacific 147
Lophodytes cucullatus. See Hooded merganser
Loxosceles apache. See Spider, brown recluse
Lugworm 62
Lumbar
 artery 161, 186
 region 198
 vertebra 160, 178, 186, 198, 200, 203, 205
Lumbricus. See Earthworm
Lumen of capillary 6
Lung 54, 75, 79, 99, 100, 107, 119, 122, 127–129, 132, 138, 142, 144, 145, 153–155, 162, 169–172, 176, 177, 185, 188, 197, 198, 215, 217, 218, 220
Lungfish 100, 107
 African 115
Lycosa. See Spider, wolf
Lymphocyte 6, 10
Lysmata debelius. See Shrimp, fire

Index

Lysmata wurdemanni. See Shrimp, peppermint
Lysosome 2, 3, 4

M

Macaw
  blue and gold 147
  scarlet 148
Macrobiotus polaris 90
Macrocystis pyrifera. See Kelp, giant
Macropus giganteus. See Kangaroo, eastern gray
Malacostraca 75, 80
Malaria 21, 24, 25
Mammalia 99, 100, 155, 156, 157, 174, 177, 192
Mammary gland 99, 155, 160, 174, 177, 181, 224
Mammillary body 194, 195, 196
Manatee 156
  West Indian 156–157
Mandible 75, 82, 88, 89, 118, 155, 160, 178, 179, 188, 200, 201, 202
Mandibular
  fossa 202
  gland 161, 163, 167
  lymph node 163
  muscle 83, 110
  notch 201
Mandrill 156–157
Mandrillus sphinx. See Mandrill
Mantis, Carolina 84
Mantle ix, 51, 53–59
Manubrium 35, 37, 200, 203
Masseter 163, 168, 181, 182, 206
Masticophis flagellum. See Snake, western coachwhip
Maturation 16
Maxilla 75, 79, 82, 87, 88, 89, 116, 117, 123, 136, 179, 200, 201, 202
Maxillary
  artery 163
  palp 87, 89
  tooth 143
  vein 130, 163
Maxillopoda 75, 76
Median
  artery 58
  eyes 79
  lobe 177
  sulcus 196
  vein 58, 186, 215
Mediastinum 198, 215, 218, 223
Medulla oblongata 114, 169, 173, 193, 194, 195, 196, 210
Medusa 33, 34, 36–39
Megaceryle alcyon. See Kingfisher, belted
Meiosis 2, 16, 21, 28
Melanerpes uropygialis. See Woodpecker, gila
Membranipora membranacea. See Bryozoan, marine
Merostomata 75, 77
Mertensia ovum. See Comb jelly, Arctic
Mesenteric
  artery 113, 130, 161, 162, 186, 189, 190, 215
  vein 164
Mesentery 122
Metacarpal 122, 123, 136, 160, 178, 200
Metamorphosis 75, 84
Metaphase 13, 14, 15
Metatarsal 122, 123, 136, 160, 178, 200
Metoicoceras geslinianum. See Ammonite
Metridium. See Sea anenome
Microscope 3
Micruroides euryxanthus. See Snake, Arizona coral
Midbrain 210
Millipede 75
  African giant 90
  American giant 76, 90
  Sonoran desert 90
Mite 75
  red 78
Mitochondria vi, 7, 21
Mitosis 2, 13, 15, 16, 28
Mnemiopsis leidyi. See Comb jelly, warty
Modicia typicalis. See Trilobite, fossil
Mollusca vii, viii, ix, 51, 52, 61, 69
Molluscs 52
Monocyte 6
Monotremes 155
Mosquito 24, 25, 75
Moth 87
  Ailanthus silkmoth 85, 87
  cynthia 84
Motor nerve 12
Mouth ix, 17, 19, 33, 35, 37, 39, 41, 43–47, 51, 53, 54, 56–58, 61, 63–65, 67, 69, 71, 75, 77, 81, 86, 87, 89, 91, 92, 94–99, 102, 105–108, 111, 119
Mucosa 220, 221
Mud puppy 122
Musa. See Banana
Muscle ix, 7, 8, 11, 20, 41, 44, 45, 48, 50, 55, 59, 61
  cardiac 12
  fiber 1
  skeletal 7, 11, 12, 69, 145, 212
  smooth 12, 217
  superficial 152, 167, 179, 180, 207, 209
Muscle tissue 1, 8, 11, 12
Muscularis 220
Mussel 51, 55
  California 52, 55
Mustelus californicus. See Shark, gray smoothhound
Mycteroperca bonaci. See Grouper
Myelin sheath 12
Mylohyoid 125, 167, 168, 181
Myocardium 216
Myotomes 19, 110
Mytilus californianus. See Mussel, California
Myxini 99, 100
Myzobella. See Leech

N

Naja. See Cobra, spitting
Narceus americanus. See Millipede, American giant
Nautilus 56, 57
Nautilus pompilius. See Nautilus
Necator americanus. See Hookworm
Neck 48, 155, 160, 163, 167, 171, 172, 174, 175, 177, 180, 181, 198, 204, 205
Necturus. See Mud puppy
Nematoda vii, viii, ix, 69, 70
Nematode 69, 70, 75
Nematomorpha vii, viii, 69
Nemertea vii, viii, 52, 61, 68, 69
Neobarrettia spinosa. See Katydid, greater arid-land
Neoceratodus forsteri. See Lungfish, African
Neornithes 145
Neospirifer. See Brachiopod, fossil
Nereis. See Sandworm
Nervous tissue 1, 8, 43
Neural tube 19
Neurofibril node 12, 211
Neuroglia 7
Neurolemmocyte 8
Neuron 7, 155, 197, 211, 213
Newt 119
  eastern 120
Nighthawk 151
Nipple 155, 181, 222, 224
Nodes of Ranvier 12
Nostril 104, 105, 106, 107, 111, 116, 118, 119, 121, 135, 140, 159, 166, 217
Notophthalmus viridescens. See Newt, eastern
Nuchal crest 179
Nuclear
  envelope vii, 1, 2, 3, 4
  membrane 1, 15, 17
Nucleic acid 1, 2
Nucleoid 2
Nucleolus 1, 2, 3, 7, 17, 211
Nucleoplasm 1
Nucleus 1–8, 12, 13, 15, 17, 21, 24, 26, 28, 211, 218
Nuttallina fluxa. See Chiton

O

Obelia 34, 36, 37
Obturator 205, 207, 208
Occipital
  artery 130, 163
  bone 200, 201, 202, 203
  condyle 131, 145, 155, 202
  lobe 193, 210
  region 173
Ochotona princeps. See Hyrax
Octopus 51, 57
  giant 56
Odocoileus hemionus. See Deer, mule
Oligochaeta 61, 64
Omentum 153, 185, 191
Omohyoid 167
Oncorhynchus keta. See Salmon, Chum
Oocyte 16, 223
Oogenesis 16
Ophiarachna incrassata. See Brittle star, green
Ophisaurus ventralis. See Lizard, eastern glass
Ophiuroidea 92
Ophisaurus ventralis. See Lizard, eastern glass
Opistophthalmus ecristatus. See Scorpion, tri-colored
Optic 20, 108, 114, 194, 195, 203, 210, 213
Oral cavity 107, 112, 119, 219
Orbicularis
  oculi 206
  oris 206
Orbit 117, 132, 179, 200
Organ 1, 8, 39, 43, 50, 51, 61, 75, 94, 99, 102, 103, 106–108, 114, 131, 137, 141–143, 153, 165, 171, 197, 198, 214, 221, 222, 224
Organ of Corti 214
Organelles vii, 1, 2, 3, 13, 21
Oriole, Bullock's 149
Orthoporus ornatus. See Millipede, Sonoran desert
Oscillatoria vi
Osculum 29–32
Osteichthyes 99, 107, 115
Osteocyte 7, 8, 11, 199
Osteon 11, 199
Ostia 29, 30, 32
Ovarian 186, 191, 223
Ovary 16, 35, 36, 45–47, 49, 64, 65, 71–73, 79, 89, 104, 118, 128, 129, 145, 162, 164, 165, 190, 212–223
Ovulation 223
Owl 146
  barn 147
  great horned 149

P

Palatine 123, 136, 202
Paleosuchus palpebrosus. See Caiman, Cuvier's dwarf
Palmaris longus 125, 182, 206, 209
Pan troglodytes. See Chimpanzee
Pandinus imperator. See Scorpion, emperor
Papillary muscle 187, 192, 216
Paramecium 22, 27, 28
Paramecium bursaria 27
Paramecium caudatum 27
Parastichopus californicus. See Sea cucumber, California
Parborlasia corrugatus. See Nemertean
Parietal 117, 143
  bone 122, 136, 179,

200, 201, 202
cell 221
lobe 193, 210
region of cerebrum 173
pleura 217
Parotid gland 161, 163, 182, 219
Passerina amoena. See Bunting, lazuli
Patella 79, 160, 178, 198, 200, 205, 206
Pavo cristatus. See Peafowl, Indian
Peafowl, Indian 149
Pectinatella magnifica. See Bryozoan, freshwater
Pectineus 167, 168, 175, 206, 208
Pectoantebrachialis 181, 182
Pectoralis 125, 152, 153, 161, 166–168, 174, 175, 179, 181–183, 206, 207, 224
Pelecanus erythrorhynchos. See Pelican, American white
Pelecanus occidentalis. See Pelican, brown
Pelican 146
  American white 148
  brown 147
Pelvic 141, 165
  brim 205
  canal 164
  cavity 198
  fin 99, 107, 109, 110, 111, 116, 117, 118
  girdle 110, 119, 122, 141, 143, 205
  vein 130
Pelvis 190, 222
Penguin 146
  Galapagos 147
Penis 45, 54, 58, 131, 164, 165, 173, 189, 190, 191, 222, 223
Peptidoglycan vii, 2
Perch 117, 118
Pericardial
  cavity 111, 198
  gland 55
  sac 197
Pericardium 56, 153, 154, 185, 215, 216
Peroneus 124, 126, 152, 166, 167, 168
Peroxisome 3
Petromyzon marinus. See Lamprey
Petromyzontida 99, 100, 104
Peucetia viridans. See

Spider, green lynx
Pharia Pyramidata. See Sea star, yellow pyramid
Pharynx 19, 41, 44–47, 63, 65–67, 71, 75, 97, 99, 102–104, 106, 108, 112, 219
Phenol wetting solution 160, 174, 177
Phidippus regius. See Spider, jumping
Phidolopora labiata. See Bryozoan, lacy
Phoca vitulina. See Seal, harbor
Phoenicopterus chilensis. See Flamingo, Chilean
Phoronida vii, viii, 51, 52, 61, 69
Phrynosoma platyrhinos. See Lizard, horned
Physalia. See Portuguese man-of-war
Pigeon 152, 153, 154
Pili 2, 159, 199
Pill bug 80
Pineal body 104, 105, 195, 210
Piranga ludoviciana. See Tanager, western
Piriformis 124, 126, 207, 208
Pituitary gland 194, 195, 196, 210, 212
Planaria 43, 44
Plasma membrane 1, 2, 3
Plasmid 2
Plasmodium 22, 24
Plasmodium vivax 25
Platyhelminthes vii, viii, ix, 43, 69
Platyzoa 43, 52
Pleural cavity 198, 217
Podiceps nigricollis. See Grebe, eared
Polycarpa aurata. See Sea squirt, ink-spot
Polychaeta 61, 63
Polyorchis penicillatus. See Jellyfish, red-eye
Polyp 33, 34, 36, 37, 40, 42
Polyplacophora 51, 53
Pons 194–196, 210
Popliteal 198, 206, 215
Porifera vii, ix, 29
Portuguese man-of-war 34, 37
Prairie dog, Utah 157
Proglottid 43, 48, 49, 50
Prokaryotic cell vii, 1, 2

Pronator teres 182, 209
Prophase 13, 14, 15
Prostate 26, 164, 165, 190, 191, 222, 223
Protoreaster. See Sea star, chocolate chip
Protozoa 1, 21, 22
Protozoan 13, 14, 21, 22, 24, 25, 26, 27, 69
Protula magnifica. See Tubeworm
Pseudocoelomates 69
Pseudotriton ruber. See Salamander, red mud
Pterobranchia 92
Pteropus hypomelanus. See Bat, Malaysian fruit
Pubis 123, 136, 152, 160, 164, 178, 200, 205, 224
Pulmonary 130, 137, 142, 145, 155, 163, 177, 186, 187, 192, 215, 216, 217, 218
Pulp 219
Pupil 213
Pylorus of stomach 200

Q

Quadratus 207–209, 221
Quail, California 146–147

R

Rabbit 174–177
  cottontail 157
  jackrabbit 156
Radius 136, 152, 160, 178, 200, 204
Rana catesbeiana. See Frog, bullfrog
Rana pipiens. See Frog, leopard
Rat 160–165
Rectum 56, 58, 67, 71, 89, 107, 142, 153, 173, 191, 219
Rectus 114, 125, 126, 161, 167, 168, 175, 183, 184, 206, 207, 208
Recurvirostra americana. See Avocet, American
Renal
  artery 161, 162, 165, 172, 173, 186, 189, 190, 191, 215, 221, 222
  capsule 222
  column 222

cortex 190, 222
medulla 190, 222
papilla 222
pelvis 190, 222
pyramid 222
vein 113, 130, 164, 170, 172, 173, 186, 189, 190, 191, 215, 221, 222
Reproductive system 155, 197, 222, 224
Reptilia 99, 100, 131, 132
Respiratory system 55, 155, 197, 217
Retina 106, 213, 214
Rhacophorus nigropalmatus. See Frog, blue-webbed flying treefrog
Rhina ancylostoma. See Guitarfish
Rhinoceros 156
  black or hook-lipped 157
Rhodactis. See Coral, mushroom
Rib 117, 118, 122, 132, 136, 143, 145, 152, 160, 178, 198, 200, 203, 218, 224
Rib cage 145, 203
Ribosome vii, 2, 3, 5
Rod 213, 214
Rodentia 156, 157
Romalea microptera. See Grasshopper, Eastern lubber
Rotifera vii, viii, ix, 69, 74
Rough endoplasmic reticulum 3, 5

S

Saccoglossus kowalevskii. See Worm, acorn
Sacroiliac 203, 205
Sacrum 160, 200, 203, 205, 208
Salamander 99, 100, 119
  red mud 120
  tiger 120
Salmon, chum 116
Samia cynthia. See Moth, cynthia
Sand dollar 91, 92
  common 93
Sandworm 61, 62, 63
Sarcoplasmic reticulum 7
Sarcopterygii 99, 107, 115
Sarcoramphus papa. See Vulture, king
Sartorius 125, 126, 152,

167, 168, 174, 175, 179, 184, 206, 208
Sauropsida 99, 131, 132
Scales 80, 88, 99, 107, 109, 116, 117, 119, 131, 135, 143
Scapula 123, 124, 136, 152, 160, 176, 178, 182, 186, 200, 204
Sceloporus. See Lizard, fence
Sceloporus magister. See Lizard, desert spiny
Sciatic
  artery 130
  nerve 185, 207, 208
  notch 205
Sclera 213, 214
Scolopendra heros. See Centipede, giant Sonoran
Scolopendra subspinipes. See Centipede, Vietnamese
Scorpion 75, 79
  Arizona hairy 80
  bark 79
  emperor 79
  flat rock 76
  tri-colored 79
Scrotum 165, 166, 191, 222, 223
Scypha 31
Scyphozoa 33, 34, 38
Sea cucumber 91, 92, 97
  California 93
  Galapagos 93
Sea nettle 39
Sea slater 80
Sea squirt, ink-spot 101
Sea star vii, ix, 17, 91, 92, 93, 94, 95
  chocolate chip 93
  yellow pyramid 93
Sea urchin vii, ix, 91, 92, 96
  green 93
  helmet 93
  pencil 96
Seal, harbor 157
Sebaceous gland 159, 199
Secretory vesicle 3
Selasphorus platycercus. See Hummingbird, broad-tailed
Semimembranosus 124, 126, 152, 167, 168, 175, 176, 184, 185, 206, 207, 208
Seminal vesicle 26, 45, 65, 66, 71, 72, 165, 222, 223

Index

Semitendinosus 126, 152, 167, 168, 184, 185, 206, 207, 208
Sensory 1, 107, 108, 119, 131, 210
  organs 99, 114, 197
  receptor 199
  tentacle 54
Sepia bandensis. See Cuttlefish
Sepiidae. See Cuttlefish
Septum 41, 65, 110, 111, 187, 192, 195, 210, 216, 217
Serratus
  anterior 180, 183, 184, 206, 207
  ventralis 166, 168, 174, 175
Sexual reproduction 13, 21, 33, 197
Shark 99, 100, 107, 109
  black-tip reef 108
  dogfish 110–113
  gray reef 108
  gray smoothhound 108
  hammerhead 101
  leopard 109
Sheep 156
  brain 193–196
  heart 192
  liver 45, 46
Shell ix, 16, 45, 46, 48, 51, 54–59, 119, 131
Shoulder 176, 182, 198
Shrimp 68, 75
  fire 80
  peppermint 76
Single-cell vi, vii, 1, 14, 21, 22
Siren, lesser 120
Siren intermedia. See Siren, lesser
Skate 99, 107
Skeletal muscle 7, 11, 12, 69, 145, 212
Skeleton ix, 33, 53, 61, 92, 99, 100, 105, 110, 117, 122, 123, 132, 136, 141, 143, 152, 160, 178, 197, 200
Skin 92, 94, 99, 107, 119, 131, 132, 140, 145, 155, 160, 165, 174, 177, 199
Skull 117, 131, 132, 136, 145, 155, 160, 174, 177, 178, 179, 201, 202
  Anapsid 132
  Diapsid 132
  Synapsid 132
Slug 51, 54
  Banana 52
Small intestine. See Intestine, small
Smooth endoplasmic reticulum 3, 5
Smooth muscle 12, 217
Snail vii, ix, 46, 48, 51, 54
Snake 100, 131, 132, 140, 143
  Arizona coral 140
  California king snake 139, 140, 144
  garter 141
  Jameson's mamba 133
  mountain kingsnake 139
  ring-neck 133
  scarlet kingsnake 140
  sea 68, 131
  western coachwhip 133
Soft palate 188, 217
Soleus 175, 185, 206
Solpugid 78
Sperm 13, 16, 29, 35, 37, 38, 64, 65, 83, 131, 197
Spermatogenesis 16
Spermatozoa 16
Spheniscus mendiculus. See Penguin, Galapagos
Sphenodon punctatus. See Tuatara
Sphenoidal sinus 202
Sphyrna tiburo. See Shark, hammerhead
Spicules 29, 30, 31, 32, 71, 72
Spider vii, ix, 75, 79
  black widow 78
  brown recluse 78
  garden 78
  green lynx 78
  jumping 78
  orb weaver 78
  wolf 78
Spine 77, 80, 91, 92, 94, 95, 96, 117, 204, 205
Spinodeltoid 179, 182, 183, 184
Spiral organ 214
Spiralia 61
Spleen 111, 118, 127, 128, 129, 137, 162, 164, 170, 171, 172, 177, 189
Sponge vii, ix, 29–32
  bath 32
  leuconoid 32
  yellow ball 32
Spongin 29
Spongocoel 29, 31
Sporulation 13
Squamosal suture 179, 201
Squid vii, ix, 51, 58, 59
Squirrel, red 101
Stagmomantis carolina. See Mantis, Carolina
Starfish. See Sea star
Sternocleidomastoid 206, 207
Sternohyoid 161, 167, 168, 175, 181
Sternomastoid 161, 167, 168, 175, 179, 181, 182, 183
Sternum 123, 145, 152, 160, 169, 178, 200, 203, 207
Stichopus fuscus. See Sea cucumber, Galapagos
Stingray
  blue-spotted 108
  round 109
Stomach 17, 54, 56, 58, 74, 79, 81, 83, 94–97, 102, 111, 118, 122, 127–129, 137, 138, 142, 144, 164, 169–172, 176, 177, 185, 189, 218–221
Stratum 159, 199
Striation 7, 8, 11
Strongylocentrotus droebachiensis. See Sea urchin, green
Styloid process 202, 204
Subclavian
  artery 112, 113, 130, 138, 153, 154, 161, 163, 171, 172, 186, 187, 188, 215, 216
  vein 113, 171, 172, 186, 188, 215
Sublingual gland 161, 219
Submandibular gland 219
Submucosa 220, 221
Subscapularis 183
Sulcus (sulci) 155, 193, 194, 196, 210
Superficial 56, 111, 114, 152, 159, 161, 166, 167, 168, 174, 175, 179, 180, 181, 184, 207, 208, 209
Superior vena cava 172, 185–189, 215, 216
Suprarenal
  artery 213, 215
  gland 213
Supraspinatus 166, 168, 175, 176, 180
Suspensory ligament 213
Suture 136, 179, 201, 202
Sweat 155, 199
Sylvilagus audubonii. See Rabbit, cottontail
Symphysis (pubis) 164, 200, 205

T

Tadpole 19
Taenia pisiformis. See Tapeworm
Taenia saginata. See Tapeworm, pork
Taeniura lymma. See Stingray, blue-spotted
Tail ix, 19, 80, 99, 102, 107, 109, 119, 122, 135, 137, 145, 150, 159, 160, 164–166, 174, 177, 223
Tamiasciurus hudsonicus. See Squirrel, red
Tanager, western 148
Tapeworm 43, 48
  pork 50
Tarantula, cobalt blue 78, 79
Tardigrada vii, viii, 69, 75, 76, 90
Tarsal bone 122, 123, 136, 160, 178, 200
Tectorial membrane 214
Telophase 13, 14, 15
Temporal 131, 132
  artery 215
  bone 179, 201, 202, 203
  lobe 210
  muscle 124
  region of cerebrum 173
Tendinous inscription 207
Tendo calcaneus 124, 125, 184, 185, 206
Tendon 12, 145, 184, 206, 207, 209
Tensor fasciae latae 166–168, 174–176, 179, 206, 208
Tentaculata 34
Tenuissimus 185
Teres 182, 209
  major 168, 175, 206, 207
  minor 206, 207
Testicular
  artery 162, 190, 223
  vein 190, 222
Testis (testes) 5, 16, 35, 36, 45–49, 64, 65, 67, 71–73, 81, 83, 104, 111, 142, 144, 164, 165, 173, 190, 191, 197, 212, 222, 223
Tetraopes tetraophthalmus. See Beetle, milkweed
Thalamus 195, 196
Thamnophis sirtalis. See Snake, garter
Thigh 159, 167, 184, 185, 198, 208
Thiothrix vi
Thoracic 175, 176
  aorta 170, 172, 189, 216
  artery 163, 171, 186
  cavity 155, 177, 197, 198
  vein 163, 171
  vertebra 160, 178, 200, 203, 218
Thorax 75, 89, 145, 160, 171, 174, 177, 180, 181, 198, 218
Thymus 169, 176, 212
Thyroid 163, 169, 170, 171, 172, 186, 212, 217
Tibia 79, 86, 88, 136, 160, 178, 200, 205
Tick 75, 78, 80
Tissues 1, 8, 10–13, 21, 29, 36, 43, 50, 51, 160, 165, 174, 177, 197, 199, 224
Toad 99, 119
  Cane or marine 121
  Colorado River 120
  Woodhouse's 120
Tongue 88, 106, 112, 118, 119, 127, 141, 166, 169, 188, 219
Tooth (teeth) 54, 96, 99, 105–109, 112, 119, 143, 155, 219
Tortoise 131
  Aldabra giant 135
  Galapagos 134
  gopher 133
Trabeculae 216
Trachea 75, 89, 137, 138, 142, 144, 153, 154, 162, 163, 171, 172, 177, 185, 187, 188, 217, 218
Trachemys scripta elegans. See Turtle, red-eared slider
Trachurus declivis. See Jack mackerel
Transverse
  colon 219, 220
  fissure 195
  groove 21, 23
  muscle 168, 183, 184, 207

septum 111
slit 132
vein 186
vertebra 123
Trapezius 163, 166, 168, 175, 176, 179, 180, 182, 183, 206, 207
Trematoda 43, 45
Triakis semifasciata. See Shark, leopard
Triceps 124–126, 152, 161, 166–168, 174, 175, 179, 180, 182–184, 206
Trichechus manatus. See Manatee, West Indian
Trichinella spiralis 73
Tridacna derasa. See Clam, giant
Trigeminal nerve 114, 194, 196
Trilobite fossil 76
Trochanter 79, 86, 88, 205
Trochlear
  nerve 114, 194, 196
  notch 204
Trochozoa 52, 61
Tuatara 131, 132
Tubercle 204, 205
Tuberosity 204, 205, 208
Tubeworm 61, 62
Tunicata 99, 102
Tunicate vii, ix, 32, 99, 100, 101, 102
Turbatrix aceti. See Eel, vinegar
Turbellaria 43, 44
Tursiops truncatus. See Dolphin, bottlenose
Turtle 100, 131, 132, 135–138
  Galapagos green sea 133
  Hawaiian green 134
  red-eared slider 134
  spiny soft-shell 133
Tympanic bulla 179
Tympanic membrane 121, 214
Tyto alba. See Owl, barn

U

Ulna 122, 136, 152, 160, 178, 200, 204
Umbilical 165, 166, 170, 171, 172, 197
Umbilicus 169, 206, 207
Ureter 144, 165, 173, 177, 189, 190, 191, 197, 221, 222
Urethra 26, 164, 165, 190, 191, 197, 221–224
Urinary 155
  bladder 118, 119, 137, 138, 164, 165, 173, 176, 177, 189, 190, 191, 197, 221, 223
  system 189, 197, 221
Urobatis halleri. See Stingray, round
Urochordata viii, 100
Urogenital
  artery 129
  opening 118, 165
  organs 165
  orifice 165, 191
  system 164, 173, 190, 191, 197
  vein 129
Ursus arctos horribilis. See Bear, grizzly
Uterine
  artery 164
  body 164, 165
  horn 164
  ostium 224
  tube 191, 222, 224
  vein 164
  wall 191
Uterus 45–49, 73, 137, 155, 165, 190, 191, 197, 222, 224

V

Vacuole 2, 21, 22, 24, 26, 27
Vagina 26, 48, 49, 54, 71, 72, 89, 164, 165, 190, 191, 222, 224
Vagus nerve 114, 171, 172, 185, 187, 189, 194, 215
Varanus exanthematicus. See Lizard, savannah monitor
Varanus komodoensis. See Komodo dragon
Varanus panoptes. See Lizard, argus monitor
Vastus 168, 174–176, 184, 185, 206, 207
Vein 58, 86, 106, 113, 127–130, 144, 161, 163, 164, 167, 170–173, 185–192, 197, 208, 215, 216, 221, 222
Vena cava 58, 128–130, 153, 161, 163, 164, 172, 173, 185–192, 215, 216, 221
Ventricle 55, 56, 106, 111, 113, 119, 122, 127, 129, 130, 137, 138, 142, 144, 154, 163, 171, 172, 177, 187, 188, 192, 195, 196, 210, 215, 216
Vermis of cerebellum 193, 210
Vertebra 99, 107, 110, 118, 122, 123, 131, 136, 143, 145, 152, 155, 160, 169, 178, 198, 200, 203, 205, 218
Vertebral 99, 100, 117, 131, 135, 136, 145, 155, 163, 169, 203, 204, 207, 215
Vertebrata viii, 99
Vessel 12, 20, 55, 63, 65, 66, 111, 164, 172, 173, 187, 192, 197, 199, 213, 216, 217, 223
Vestibular
  gland 164, 165
  membrane 214
  window 214
Vestibulocochlear nerve 194, 214
Viscera 118, 128, 129, 137, 154, 160, 162, 164, 165, 170, 174, 176, 177, 198, 221
Vitreous humor 213
Volvox vi
Vulture, king 148

W

Water bear vii, 75
Water flea 80
Woodpecker, gila 151
Worm
  acorn vii, 92, 98
  bloodworm 62
  earthworm 61, 62, 64, 65, 66
  eye worm 70
  flatworm vii, ix, 13, 14, 43
  filarial 69
  gill-wing 92
  goblet vii
  heartworm 73
  hookworm 69, 70
  horsehair vii
  horseshoe vii
  lugworm 62
  milky ribbon 68
  pinworm 69, 70
  proboscis vii, 61
  roundworm vii, ix, 69, 70
  sandworm 61, 62, 63
  segmented vii, ix, 61
  spiny-crown vii
  tapeworm 43, 48, 50
  trichina 69
  tubeworm 61, 62
  zebra ribbon 68
Wrist 160, 166, 174, 177
Wuchereria bancrofti. See Nematode

X

Xenia 42
Xenia umbellata. See Xenia
Xiphihumeralis 179, 183
Xiphoid process 203

Z

Z line 7
Zenaida macroura. See Dove, mourning
Zoecium 60
Zygomatic
  arch 179, 202
  bone 179, 200–202
Zygote 13, 16, 29, 37, 49